D1237327

A Field Guide to Southern Mushrooms

Nancy Smith Weber and Alexander H. Smith

Photographs by Dan Guravich

Ann Arbor
The University of Michigan Press

Copyright © by The University of Michigan 1985
All rights reserved
Published in the United States of America by
The University of Michigan Press and simultaneously
in Rexdale, Canada, by John Wiley & Sons Canada, Limited
Typeset in the United States of America.
Printed and bound in Japan.

1988   1987   1986   1985   4   3   2   1

Library of Congress Cataloging in Publication Data

Weber, Nancy S.
  A field guide to southern mushrooms.

  Bibliography: p.
  Includes index.
  1. Mushrooms—Southern States—Identification.
I. Smith, Alexander Hanchett, 1904– .   II. Guravich,
Dan.   III. Title.
QK605.5.S68W43   1985   589.2′0975   83-23553
ISBN 0-472-85615-4

Preface

The only thing required is to know each of them [edible mushrooms] well, and to be always on the lookout that no bad stranger be mixed up with them.

Auguste Barthélemy Langlois

This advice to potential mycophagists is as valid today as it was in 1898 (Langlois 1900, p. 14). Our aim is to help the user "know each of them well" and avoid "bad strangers." To this end a selection of distinctive southern mushrooms is presented here. Species that are good to eat, common and conspicuous, or poisonous are emphasized. Species that illustrate diversity in the fungi, are special challenges to the collector, or that are otherwise intriguing complete the selection.

Becoming acquainted with southern mushrooms is a formidable task. More kinds of mushrooms are thought to occur in the southeastern United States than in any other region of comparable size in North America. The multiplicity of habitats, diversity of vascular plants, long growing season, and moist, relatively warm climate provide a wealth of opportunities for fungi. Between 3,000 and 5,000 kinds of mushrooms are estimated to occur in the South. There is, however, no reasonably complete treatment of southern mushrooms as yet. Many parts of the South remain to be explored for mushrooms, and many large groups of mushrooms have not been studied in detail. Nevertheless, there are a number of distinctive species, both poisonous and edible, that can be identified, and some eaten and enjoyed, by anyone with an interest in wild things.

The diversity of southern fungi, particularly mushrooms, has been a source of fascination for residents and visitors for over a century and a half. Studies on southern fungi as an independent discipline appeared as early as the first quarter of the nineteenth century. By the middle of the century, the South was the center of North American mycology. The Civil War and its aftermath halted scientific studies on southern fungi for many years. Only after the turn of the century was there a resurgence of the strong tradition of southern mycology. In recent years there has been an increased interest in both the popular and scientific aspects of southern mushrooms.

Acknowledgments

Dan Guravich originated the idea of the project when he found he could not identify many of the mushrooms in and around Greenville, Mississippi. He undertook the tasks of collecting and photographing the mushrooms of the region. His specimens were studied by Weber and Smith and deposited in the Herbarium of the University of Michigan. They serve to document most of the photographs in the book. Several joint collecting trips to other areas were made to gather additional information, specimens, and photographs. For each species, the number of the collection used for the illustration is listed, when known, in appendix 1.

Many individuals, institutions, and groups aided us. The Herbarium of the University of Michigan, directed by Dr. Robert Shaffer, provided working space for the authors and the repository for the collections. We were privileged to use the facilities of the Highlands Biological Station, Highlands, North Carolina; the Gulf Coast Research Laboratory, Ocean Springs, Mississippi; and the Harrison Experimental Forest of the De Soto National Forest with headquarters at Gulfport, Mississippi. Many collections and photographs were made along the Natchez Trace Parkway in Mississippi, and we are pleased to acknowledge the help of the Parkway staff. The North American Mycological Association allowed us to participate in two forays near Tuxedo, North Carolina. Aileen Stanley and Dr. and Mrs. Bedford Floyd provided significant advice and help. The National Space Technology Laboratory (NSTL) nature trail at Bay Saint Louis, Mississippi, was opened to us. Dr. William Cibula of the NSTL, a mycologist himself, generously shared his knowledge of the mushrooms and mushrooming along the Gulf of Mexico. Several professional mycologists—Joseph Ammirati, Howard Bigelow, Kenneth Harrison, Currie Marr, Clark Ovrebo, W. W. Patrick, Jr., and Robert Shaffer—examined specimens and shared their opinions with us. The final determinations were made by the authors.

Joy Guravich, Helen Smith, and James Weber each contributed their support and encouragement to the photographer and the authors throughout the project.

Contents

Introduction

We have been asked many times what we mean by "southern mushrooms." It is not an easy question to answer, and in any case the mushrooms will not read our definition and abide by it. The region in which this book will be most useful is the southeastern United States. Although we can indicate the boundaries of this area, local deviations are numerous. Furthermore, many of the mushrooms that occur in the South may be found in other parts of this continent and even on other continents. We have defined the "South" on general patterns of vascular plant distribution, geology, and climate—all of which influence the diversity of mushrooms. The northern boundary extends roughly westward from the Atlantic Ocean along the southern edge of Pennsylvania, then along the Ohio River to its junction with the Mississippi, and westward to the eastern edge of the Great Plains. The western boundary is the eastern edge of the Great Plains. The Gulf of Mexico and Atlantic Ocean form the remaining boundaries. Our studies were concentrated in the Appalachian highlands, Mississippi Delta, and coastal plains along the Gulf of Mexico and Atlantic Ocean—areas where not only are there a lot of mushrooms but also many mushroom hunters.

The kinds of mushrooms that may be found in the South present intriguing questions to anyone interested in the distribution of fungi. Some species are basically the weeds of the mushroom world that are widely distributed and common. Some are endemic to the South, i.e., known from no other region. Others are at the northern limit of their range along the Gulf of Mexico; while in the Appalachian Mountains some species are at the southern edge of their range (see appendix 2). Furthermore, many species described from the South occur in such distant regions as Africa, Sri Lanka, Papua New Guinea, and Japan. As you travel, be it through books and pictures or actual trips, you may encounter familiar southern mushrooms in exotic places and wonder, along with the scientists, how they came to be where they are.

How to Use This Book

Before you take your first bite of a wild mushroom, take time to learn about mushrooms: what their place and function in the natural world is; how to collect, identify, and prepare them; and what the pitfalls are. Skim through the book, glance at the photographs to obtain an idea of the variations in shape, color, and form that you may encounter. Then read and study the introductory materials, especially the section on "From Field to Frying Pan." Discussions of mushroom nomenclature and the history of mycology in the South can add new dimensions to your knowledge.

The process of identifying an unfamiliar mushroom is one of trying to match your specimens with an illustration and/or description of a known species. While searching for a good fit, you observe the features of your specimens and those of known species and become familiar with the characters used to identify mushrooms. At first you may prefer to leaf through the pictures and compare the specimens with them, but you will soon discover that individual specimens are not all exactly alike, and questions arise about the significance of the differences. Now is the time to learn how to use the keys to sort through the possibilities and then refer to the photographs and descriptions. In this way you learn how to distinguish one kind of mushroom from another as well as identifying the unfamiliar mushrooms.

Keys are devices to help one sort possibilities. To use them, read the first pair of choices in the first key in the book and decide which one best describes your specimens; scan to the right of the page and note the number given there. A choice in a key is called a lead and the number to the right refers to another pair of leads. Go directly to that pair of leads, skipping all intervening ones, read that pair and again choose the one that best describes the unfamiliar mushroom. Continue in this manner until you arrive at the name of a group, genus, or species. You may have to work through several keys before arriving at the name of a species. Once you do, turn to that description and photograph and compare your material with them; if all the important characters match, the identification is probably correct.

In the descriptions of individual species the important field characters are emphasized in the first section; they are part of the "essence" of the species. Data on microscopic features are included for those who have access to a microscope. Techniques for studying these characters are discussed in some of the references listed in the bibliography. The information on edibility for each species indicates only what has been reported about a species or experienced by us. There is no guarantee that just because a species is listed as edible you can eat it without trouble. Personal reactions vary to foods, and mushrooms are no exception. Neither the authors nor the publisher accept responsibility for any of the identifications made by users of this book or for the consequences of eating any of the mushrooms treated here.

Species are grouped in the text in much the same manner as they are by scientists in technical works. This arrangement facilitates making comparisons between similar species and helps the user develop an "eye" for various groups. When you consult advanced books on mushrooms, some familiarity with these groups will be helpful. Furthermore, once you can recognize certain of them, the next time you collect a member of a group you can go directly to the appropriate section of the book.

Because no book includes all the mushrooms known to occur in the South, or anywhere else for that matter, problems and questions are inevitable. Help can often be obtained from the botany, plant pathology, or biology departments of local colleges and universities. The experts need your help in the form of a description of the unknown mushroom that includes features such as color, presence of slime on the cap, odor and taste of the raw flesh (chew and spit, do not swallow) that cannot be learned from dried specimens. Then dry your specimens as described on p. 9. Together with the description, they form a unit that can be studied and often identified.

Mushrooms: Function and Structure

Few people realize when they look at a mushroom that it is only a portion of the fungous organism. The mycelium, also called the

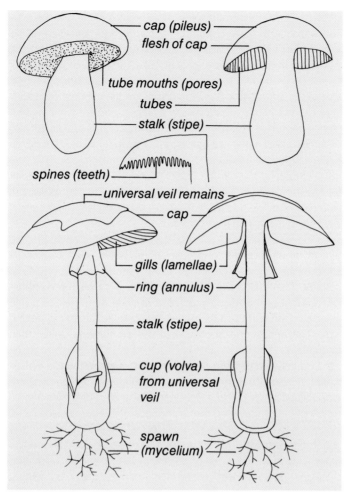

cap (pileus)
flesh of cap
tube mouths (pores)
tubes
stalk (stipe)
spines (teeth)
universal veil remains
cap
gills (lamellae)
ring (annulus)
stalk (stipe)
cup (volva) from universal veil
spawn (mycelium)

spawn, that produced the mushroom is seldom visible. It obtains nutrients and water from its surroundings and consists of a mass of threadlike structures called hyphae. Individual hyphae are too narrow to see without a microscope; however, strands of them can often be found in old logs, leaf piles, and the soil. Mushrooms themselves are also composed of hyphae which are usually highly modified. The familiar structures commonly called mushrooms are just one phase in the life of a fungus, the reproductive phase. On or in the mushroom, spores are produced that are the bridge between generations. The spores perpetuate and disperse the species. Mushrooms and their spores are somewhat similar in function to the flowers, fruits, and seeds of vascular plants; however, mushrooms are not true fruits. Instead of calling them fruits, the terms fruiting body or fruit body are used to refer to what we usually call a mushroom when we want to distinguish it from the fungous organism which produced it. Both the fruiting body itself and the entire organism are called mushrooms on occasion; however, one can usually tell from the context which meaning is involved.

The variations in shape, form, and structure of fruiting bodies reflect different approaches to producing, liberating, and dispersing spores. The most common type of fruiting body consists of a cap, stalk, and gills. The gills are thin plates of tissue that resemble the pages of a book held by its spine. The surfaces of the gills are covered with one or more types of specialized cells. Basidia are the cells on the gills on which the spores are produced. When the spores are mature they are forcibly flicked off the basidia then fall toward the ground from between the gills. If no stalk were present, most of the spores would pile up just below the cap that produced them and not be dispersed. When a stalk is present and elevates the hymenophore (the spore-producing region), the spores fall into air currents and many are carried away by them to new localities.

All mushrooms that bear their spores on basidia belong to the group of fungi known as the Basidiomycotina. Not all members of the Basidiomycotina have this familiar type of fruiting body, but each type is efficient in its own way in producing large numbers of spores and facilitating their dispersal.

A minority of the fungi with conspicuous fruiting bodies that we refer to as mushrooms form their spores within specialized cells called asci and shoot the spores away from the fruiting bodies at maturity. They belong to the group of fungi called the Ascomycotina. Most of the Ascomycotina we include bear their asci and associated structures on the upper, exposed surface of the fruiting body. The spores are shot up into the air currents rather than dropped into them.

Individually, spores are too small to be seen without a microscope; in mass they resemble fine dusting powder. Each spore consists of a single cell and the range in size is roughly from 3 to 30 micrometers (abbreviated μm) in the mushrooms we include here. If you wish to obtain spores for study, cut off the stalk from a mature mushroom with gills, place the cap gills down on a piece of white paper, wrap in waxed paper or cover with a bowl, and do not disturb for several hours. When the cap is lifted off the paper, a powdery deposit of spores can be seen. Depending on what species of mushroom was used, the powder may be white, yellow, some shade of red, brown, or black. The color of the spores in mass in a spore deposit (or spore print) is one character used to group different kinds of mushrooms.

Spore dispersal is usually a cooperative venture involving the fruiting body and the environment. Air currents are the primary agents of long distance spore dispersal. Some other methods of dispersing spores will be discussed in connection with various groups of mushrooms.

Once a spore has come to rest, it may die, remain dormant until conditions change, or germinate, i.e., grow. Only a few spores out of the millions that may be produced by a single fruiting body will land in places where they can germinate and grow into a mycelium. A germinating spore sends out one or more hyphae which branch and grow to form a mycelium. Two mycelia of the same species but produced by different spores must come into contact to form a mycelium capable of producing fertile fruiting bodies.

The hyphae of a mycelium obtain water and nutrients from their surroundings which are used in the growth and development of the organism. Many fungi are saprophytes, extracting nutrients from dead organic matter. Some are parasites, using living plants, animals, or other fungi as their host. Many mushrooms participate in mycorrhizal associations in which the roots of vascular plants, particularly trees and shrubs, and the fungous mycelium are intimately associated. The trees may receive some minerals and water that the fungi take up from the soil, and the fungi receive various substances from the trees. The association of certain species of mushroom with particular kinds of trees is so predictable that a mushroom hunter looking for a particular kind of mushroom should first locate the appropriate tree then look for the mushroom. Trees are easier for most of us to find than mushrooms.

Before a mycelium can produce fruiting bodies, a number of requirements must be met. They differ from species to species but common variables include temperature, moisture, nutrients, and seasonal weather patterns. A majority of species, including many good edible ones, have not been successfully cultivated. The only way we can enjoy them is to collect them in their natural habitats.

In summary, when we pick a mushroom we are taking away the reproductive phase of the mushroom organism. The mycelium usually persists and may produce more mushrooms that season or another time. However, if, as is the case of chanterelle picking in parts of Europe, most of the mushrooms are harvested before they mature and do not release spores, the "spore rain" of the species is reduced which in turn lessens the likelihood that new mycelia will become established. Over a long period of years old mycelia will die out and no new ones will be established. While it is conceivable that the crop of wild mushrooms might be diminished in this way, there is little likelihood of that happening in the foreseeable future in the South.

Mushrooms: From Field to Frying Pan

Fruiting bodies may be ephemeral, lasting only a few hours; short-lived, functioning for a few days; or long-lived, persisting for weeks, months, or even years. Most of the edible species belong in the second category. The basic equipment for collecting edible mushrooms includes an open basket for carrying collections, a knife or trowel for detaching them from their substrate, and waxed paper in which to wrap them. When picking mushrooms, carefully collect each specimen—being sure to include the base of the stalk. Place all the specimens of the same kind from the same patch (presumably the product of one mycelium) in a square or rectangle of waxed paper, roll the paper to form a rough cylinder, and twist the ends to form a package. In the South especially, plastic wraps and plastic bags should be avoided. Plastic holds moisture too well and allows heat to build up. As a result mushrooms wrapped in it often "stew in their own juices" in warm or hot weather. Specimens handled in this manner should not be eaten. When you find specimens you want to preserve for later use, pick them carefully as always, check their characters, be sure of your identification before trimming away any dirt or part, then wrap the cleaned mushrooms or place them in a paper bag. It is easier to pick and keep mushrooms clean than it is to clean them later. Keep each species separate and save some intact specimens of each kind in case they are needed for further study.

In addition to the basic equipment, a hand lens for looking at details; a magnetic compass (and knowledge of how to use it); a map of the area where you are collecting; a container of water; and a whistle (in case you get lost and want to be found) are useful. No one should go into the woods without some awareness of the common hazards they might encounter and some idea of how to avoid and/or deal with them. Mushroom hunters may become thirsty or even dehydrated in hot weather. They may encounter plants that can cause rashes and skin irritations, mosquitoes, chiggers, fire ants, poisonous snakes and spiders, and the occasional alligator—any of which can take the fun out of collecting.

Careful collecting should be followed by careful transporting so that the specimens arrive home in good condition. During the hot summer months, mushrooms deteriorate in a few hours after being picked unless they are kept cool. An air-conditioned car and house or laboratory are useful—both the hunter and the hunted benefit from cooling off. Some hunters take ice chests with packaged ice on their trips to keep their collections cool on long drives. Many go collecting early in the day and plan to be home before the specimens—or the hunters—become overheated.

Identifying specimens accurately is extremely important for anyone who plans to eat wild mushrooms. The objective is to match individual specimens with the description and/or illustration of a species. The description of a species is "distilled" from the examination of many individuals so you must learn to distinguish which characters are important in placing a specimen in a species and which merely represent individual variations. For example, the presence of gills instead of tubes on the underside of the cap is likely to be an inherited feature, stable from generation to generation and important in making an identification. In contrast, the exact dimensions of each fruiting body are likely to vary from individual to individual within limits characteristic of the species. Your specimens will seldom, if ever, exactly match those in an illustration, but by comparing the characters emphasized in the descriptions with those of your specimen and the illustration, you can identify individual specimens with reasonable accuracy. Only specimens whose identity you are sure of and that belong to species known to be edible should be eaten. A bit of caution in this phase of mushroom hunting can save a lot of trouble later.

Eating wild mushrooms can be safe and enjoyable if certain precautions are observed; most of them are a matter of common sense. Chief among the cautions is to know what you are eating and eat only species of proven edibility. Try only one new kind of mushroom a day; some types of poisoning are slow to appear. Carefully check the identity of each specimen before trimming off any parts. Eat only fresh, worm-free specimens that show no signs of spoilage. To check for worms (actually the larvae of insects), cut the stalk off crosswise or slice the fruiting body in half longitudinally and look for tiny tunnels or small white worms. If either sign is present, trim until all the wormy part is gone or throw out the entire specimen. Be moderate in the size of serving you eat. Mushrooms are not readily digested and if consumed in quantity may cause indigestion or stimulate the gastrointestinal tract much as do some laxatives. Just as some people cannot tolerate certain foods, some cannot tolerate mushrooms that others eat and enjoy. Because such reactions cannot be predicted on an individual basis, each person should try each species in a small quantity to check for such idiosyncracies. No more than a tablespoon of cooked mushrooms for a child or a quarter of a cup for an adult should be eaten. Unless stated otherwise, all our edibility ratings are for cooked mushrooms. Do not expect mushrooms that are edible when cooked to be safe when eaten raw—some are and some are not. With these guidelines in mind, if you select the species to try from among those rated as good edible species, accurately identify your specimens, and eat only fresh, well-cooked specimens, the chances of being poisoned by mushrooms are minimal. Remember, however, that the decision to eat or not to eat is yours.

Mushrooms can be preserved in a number of ways. In all cases, use firm, fresh, worm-free specimens whose identity was checked carefully and process them at the peak of their goodness. Many people lightly sauté mushrooms, divide them into meal-sized packages and freeze them. The mushrooms can be thawed and used in soups, casseroles, gravies, or other dishes. Our favorite way is to slice the cleaned mushrooms into thin pieces then dry them. A home food dehydrator, a set of screens over a hot plate, or a hair dryer may be used. Only if the weather is sunny and dry is sun drying reliable—and such is not likely in good "mushroom weather." The mushrooms should be crisp like a fresh potato chip when they are dry. They should dry slowly in twelve to eighteen hours rather than being overheated, in which case they break down into a tarlike mess. We package our dried mushrooms in plastic bags and store them in a freezer; they may also be kept in tightly closed jars. Soak the dried mushrooms in warm water until they reach the desired consistency or according to the recipe directions and then cook them. Pickling is popular in some areas. For this method, species with firm fruiting bodies and no slime layers are recommended. Canned mushrooms should be processed under pressure to reduce the chance that the organisms responsible for causing botulism will develop.

Poisonings from eating wild mushrooms can take many forms, from mild indigestion to fatal illnesses, although there are relatively few of the latter. One of the main functions of a guide such as this is to relate what is known not only about the edibility of various species but also their potential toxicity in the hope that the users will avoid seriously poisoning themselves and others. Reports of different adverse reactions appear every year, but at present the majority are caused by seven types of poisoning.

If mushroom poisoning is suspected, there are certain things that should be done no matter what toxin is involved. Contact a physician and/or poison control center and have it determined whether the illness is actually mushroom poisoning. Let them make the decision. If mushroom poisoning is diagnosed, usually the first step in treatment is to remove any undigested mushrooms from the victim. Current advice to physicians treating cases of mushroom poisoning is to treat the symptoms expressed by the patient. It is of secondary importance in treatment to identify the mushroom that caused the illness, but having that information may help the physician anticipate the course of the illness. Mycophagists can serve science and their fellow mycophagists by always keeping some uncooked specimens on hand when they experiment with a new species so that if they are poisoned, the cause can be established. It has been shown that the sooner treatment is initiated after the mushrooms are eaten, generally the faster and more complete the recovery. The following discussion of the various types of mushroom poisoning (perhaps better called intoxications) is superficial. More extensive discussion of the subject may be found in the books listed in the section of suggested references.

Gastrointestinal upsets and individual idiosyncracies. The bulk of the problems in this group are disturbances of the digestive system. In most healthy adults who do not become dehydrated, this type of problem is generally self-limiting, but again medical advice should be sought. The compounds that cause such upsets are not known in most cases. Almost every species that has been rated edible, as well as many others, has at one time or another caused digestive system upsets. However, several species consistently cause such reactions in most or all people who eat them, and are thus best avoided. Several species of *Russula*, *Scleroderma*, *Agaricus*, many species of *Boletus* in which the flesh stains blue when bruised, *Chlorophyllum molybdites*, and *Omphalotus illudens* are known to cause such problems when eaten raw or cooked. Additional species that are generally rated edible but that may cause gastro-intestinal upsets when eaten raw or insufficiently cooked are *Laetiporus sulphureus* and both species of *Armillariella*. A second type of reaction is that of individual idiosyncratic reactions such as allergies to specific fungi. These cannot be predicted. It is the possibility of experiencing such reactions that makes it imperative that each person try each kind of mushroom for himself or herself.

Coprine poisoning. Only if one drinks a significant quantity of alcohol a few hours to perhaps a day or two after consuming a fungus that contains coprine will this type of poisoning appear. Coprine may persist in the body several days. Symptoms appear within an hour of drinking alcohol and include flushing of the face and neck, a metallic taste in the mouth, nausea, vomiting, and sweating. Recovery is usually spontaneous and occurs within a few hours. Two species that cause this type of reaction are *Coprinus atramentarius* and *Clitocybe clavipes*.

Muscarine poisoning. This type of intoxication is characterized by the PSL combination of symptoms: profuse perspiration, salivation, and lachrymation. They may be accompanied by diarrhea, low blood pressure, small pupils, and difficulty in breathing. The symptoms develop from fifteen minutes to two hours after eating the mushrooms. In severe cases physicians may administer atropine as an antidote. Many species of *Inocybe* have been shown to contain significant amounts of muscarine as does *Clitocybe dealbata*. Muscarine occurs but is not the principal toxin in *Amanita muscaria* and is suspected to occur in some species of *Boletus*.

Ibotenic acid—Muscimol poisoning. Ibotenic acid is present in some mushrooms and is quickly converted in the body to muscimol, a more potent compound. These substances cause dizziness, increasing incoordination sometimes accompanied by altered perceptions of the surroundings, and visions within a couple of hours after ingesting the mushrooms. The reaction ends with a period of deep sleep. Some people seek these sensations deliberately and use mushrooms that contain ibotenic acid "recreationally," but their use is not without danger. Some "trippers" have nearly choked on the copious mucus they produced but could not clear from their throats. *Amanita muscaria* is the best-known of these fungi.

Psilocybin and psilocin poisoning. These compounds, like ibotenic acid and muscimol, exert their greatest effect on the central nervous system. Psilocybin is chemically related to LSD and, like it, may cause hallucinations and drowsiness. Some people experience residual headaches after using mushrooms with these compounds. In people who are not expecting these symptoms they can be very disturbing; others seek them deliberately, and one's attitude is said to have considerable influence on the experience. The use of psilocybin-containing mushrooms is not without risk. Some people seem to develop violent, sometimes fatal, allergic reactions to these fungi. It is illegal to possess, sell, or transport these fungi. *Psilocybe cubensis* is the most famous of the psilocybin-containing mushrooms.

Monomethylhydrazine poisoning. None of the species included in this book have been proven, to our knowledge, to release this compound in quantities sufficient to cause illness. Monomethylhydrazine is released during the cooking of fruiting bodies of certain species of *Gyromitra* especially *G. esculenta* and *G. infula*. It is also a component of some rocket fuels. The symptoms of this type of poisoning appear several hours (eight to twelve) after ingestion and include nausea, vomiting, diarrhea, cramps, and may be followed by signs of liver malfunction and even death. Children seem to be the most likely to suffer serious consequences in this type of poisoning. Good supportive care is generally adequate to insure recovery.

Amanitin poisoning. Of all the types of mushroom poisoning this is the most serious. It has been estimated that about 90 percent of the fatalities caused by mushrooms are traceable to species that contain amanitin. Among the species known to contain dangerous quantities of amanitin are *Amanita bisporigera*, *A. virosa*, *A. phalloides*, and *Galerina marginata* and its relatives. This toxin interferes with the basic workings of cells, especially those of the liver. The effects of amanitin on the liver can be detected in blood samples almost immediately after the fungus is eaten but no symptoms are apparent until about twelve to thirty hours later. The first symptoms to appear consist of nausea, vomiting, and bloody diarrhea. They abate but some hours to a day or so later, there is clear evidence of liver malfunction and death may occur in four to six days. Data gathered on this type of poisoning in Europe have shown that the sooner treatment is started, preferably within a few hours, the better the chances for recovery. If treatment is started soon, most victims will recover. There is no general agreement on how to treat cases of poisoning by amanitin, other than to treat the symptoms of the victim. As yet there is no sure cure.

Mushrooms: Nomenclature

The importance and utility of names was expressed by Auguste Barthélemy Langlois almost a century ago:

> There is much in a name. It is the synthesis of all that is known of a thing or person named. If the beasts of the earth and the fowls of the air have names, so the plants, also living creatures, must have names by which they may be recognized with all their peculiarities and qualities. Therefore, it is of the greatest importance to know the accepted names of plants, and, by their names, their situation and relation in the vegetable kingdom. [1896, p. 5]

The names Langlois mentioned are the scientific names which provide a universally accepted means of referring to organisms. In the binomial system for scientific names introduced by Carl Linnaeus (1707–78), each species has one and only one correct name. It consists of two words: the name of the genus followed by the specific epithet. The names are in Latin, or treated as if they were Latin, because Latin is a dead language and does not change. Many names emphasize some feature of the organism that can be ascertained by breaking the code—translating the Latin. Decoding scientific names can be done with the aid of a good dictionary; as your understanding of them grows, so will their usefulness.

Some people are intimidated by scientific names at first and want to know the "common names" of the mushrooms they find. Their wishes can be satisfied only in part. Whereas most conspicuous plants and animals have one or more common names, relatively few North American mushrooms have attracted enough attention to have received even one common name (i.e., a vernacular name that has been used by many people for many years). Whereas common names are not efficient means of communicating about species across cultural and/or linguistic boundaries, the scientific name is the same in all countries. It is a name that is understood wherever mushroom hunters gather. Mushroom hunting is one pastime where the names that are truly common are the scientific names. Anyone with a serious interest in the subject should learn them first.

Scientific names are formed, applied, and used according to a system of rules which for fungi are set forth in the *International Code of Botanical Nomenclature* (Voss et al., 1983). One of the basic principles of this code is that names are based on the "type method." The type of the name of a species is a specimen that is designated as the type collection (or holotype), which is preserved in a herbarium where qualified investigators may examine it. A significant consequence of this method is that the types of two or more names can be compared. If they represent a single species, then the oldest name that is in accordance with the rules should be used for the species. There are no type collections for the names of many species described before the early part of this century, a situation which makes it difficult to fix the application of those names in many cases. Substitute types have been selected for some of these names, but in cases where the original description is not particularly detailed and there are no types, problems arise in deciding which of the variants that fit the description the describing author actually had before him. We use the term "collective species" to indicate a situation where several variants with similar macroscopic features are included under a single name, e.g., *Armillariella tabescens* and *Pleurotus ostreatus*.

A second principle of the *Code* is that the earliest name that is in accordance with the rules of nomenclature should be used for each species. Starting points for the naming of species have been designated. The *Species Plantarum* by Linnaeus, which is treated as if it were published 1 May 1753, is the starting point for names of fungi. There are some important exceptions to this rule. One group of exceptions includes all rusts, smuts, and gasteromycetes treated in the *Synopsis Methodica Fungorum* by Christiaan Hendrick Persoon (1761–1836). The other group includes miscellaneous fungi (most of the fungi treated in this book are in this group) described in the three volumes of the *Systema Mycologicum* and the *Elenchus Fungorum* by Elias Magnus Fries (1794–1878). Names of species that are treated by Persoon or Fries take priority over the same names if used for the same or a different fungus by an earlier author, i.e., they have a special status. For purposes of nomenclature, the *Synopsis* is considered to have been published on 31 December 1801 and volume one of the *Systema* on 1 January 1821.

The name(s) of the person(s) who first validly published the name of a species may be cited after the name of the species, e.g., *Lactarius corrugis* Peck. The name of the authority may be abbreviated (a list of abbreviations used in this book is given in appendix 3). From such a citation we know that Peck described this species and placed it in the genus *Lactarius* where the authors of the present work believe it still belongs. Several other types of citations are common and each indicates something about the history of the name. Sometimes more than one person may be cited, an indication they jointly proposed the name, e.g., *Boletus piedmontensis* Grand & Smith. In the case of *Tylopilus alboater* (Schweinitz) Murrill, Schweinitz originally described a species with the specific epithet "alboater" but he did not place it in the genus *Tylopilus*. Murrill transferred Schweinitz's species from its original genus, in this example *Boletus*, to the genus *Tylopilus* at a later date. In the example of *Boletus griseus* Frost in Peck, Frost supplied the description of the fungus and Peck was the author of the publication in which it appeared. The inclusion of Peck's name in the citation is optional but useful. Our final example, *Gyromitra caroliniana* (Bosc : Fries) Fries, is more complicated. Bosc proposed the name *Morchella caroliniana* for a species he described in 1811. Fries adopted that name in the *Systema*, thus by current rules it has a special status. Years later Fries transferred the species from *Morchella* to *Gyromitra*; consequently his name appears a second time in the citation.

We have tried to follow the 1983 *Code* in citing both the describing and sanctioning author(s) for older names of species. Many of the provisions of this *Code*, however, have not been "use tested" and are creating controversy among mycologists. Under the 1983 *Code*, the names *Lactarius* and *Cortinarius* become *Lactaria* and *Cortinaria*, and the names of some genera will either have to be conserved or changed completely. Because a field guide is no place to introduce untested changes, we have followed the previous edition of the *Code* in selecting names for genera. Nomenclature is often as confusing as taxonomy.

The people mentioned so far all had an impact, direct or indirect, on southern mycology. A few of them, and some other men and women who dealt directly with mushrooms in the South, are introduced in the next section. Sometimes they are as interesting as the fungi they found.

Students of Southern Mushrooms

I look upon our work at present in this light.—Here is a most untrod, almost unexplored wilderness of new and strange forms—a field white for the harvest. Shall we stroll over these grounds, gather what we can, and arrange and assort as best we may, leaving to those who come after us the task of rearranging and (doubtless) of correcting our many errors—or shall we occupy only a little corner with more close inspection, and leave the great field beyond entirely out of sight?

William Henry Ravenel

The dilemma expressed by Ravenel (in Stevens 1932, p. 141) is one that has faced students of natural history from early times to the present and will persist into the future. Although Ravenel viewed himself as a pioneer in American mycology, he was far from the first person whose observations on North American fungi found their way into print. The history of studies on southern fungi cannot be told in full here. Rather, we will introduce some of the men and women who made important or interesting contributions to the subject in the past and whose names are often cited in conjunction with those of southern mushrooms.

Certain problems confronting early students of fungi have not yet been completely solved. One problem, as Underwood and Earle stated, is that "The specimens . . . are the favorite food of certain insect larvae, and if left over night will often be found to have changed into disgusting heaps of corruption by morning" (Underwood and Earle 1897, p. 274). While students of vascular plants could press and dry their collections with reasonable ease, mushrooms posed special problems because of their high moisture content and tendency to rot rapidly or be eaten. Some collectors pressed their fungi, including mushrooms and puffballs; the results are seldom pleasing or informative. Other methods of preparing specimens included stringing up the specimens in a dry place, putting them in wire baskets over a source of hot air, or just setting them on a table in the hope that they would dry before they rotted or were eaten. Dried mushrooms were often moistened, then pressed and glued to paper or cardboard. Some specimens were dipped in or sprayed with poison in order to kill insects, thus further obscuring the details of the fruiting bodies. In the period when roads were poor or nonexistent and travel was often by horse, the fragility of mushrooms and their ephemeral nature posed additional complications. It is hardly surprising that studies on the mushrooms of the South lagged behind those on vascular plants.

Before 1811 published accounts of southern fungi were usually in the form of isolated observations by travelers, explorers, or as a small part of a catalog of the local flora. The comments by John Banister (1650–92) and Thomas Walter (1740–89) were typical of the early period. Both men were born in England. Banister was a clergyman who settled in Virginia in 1678 and spent much time studying the natural history of Virginia. Walter settled in South Carolina by 1769 and was a planter along the Santee River during the Revolutionary War. His observations on the flora of that region included notes on twenty-one fungi.

With the appearance of a paper titled "Mémoire sur quelques espèces des Champignons des parties méridionales de l'Amérique septentrionale" by Louis Augustin Guillaume Bosc in 1811, studies on southern fungi entered a new era. Bosc (1759–1828) was born and educated in France. In 1796 he sailed for Charleston, South Carolina, partly to put distance between himself and the dregs of an unfortunate love affair. He stayed in North America only a few years before returning to France. Bosc was known for his studies on and/or interests in several aspects of zoology. The 1811 paper is apparently his only contribution to American mycology. In it he described fourteen species and one new genus of fungi, limiting himself to those he was sure were undescribed. Many of these species still bear the names he gave them.

From late in the first quarter of the nineteenth century until after the middle of the century, the South was the center of mycological activity in North America. During this period, for the first time, persons born in North America rather than abroad made important contributions to the subject. Whereas most of the early reports on southern fungi were published in Europe, by the end of the century most were issued in the United States.

One famous publication on the fungi of North Carolina was published without the author's knowledge. It seems that Lewis David de Schweinitz (1780–1834) took a list of fungi he had collected in North Carolina to Europe in 1818. There he gave the list to a friend who was active in a natural history society in Leipzig. Unknown to Schweinitz until he received the reprints, the article was published in the society's journal four years later. He was surprised but pleased. Schweinitz was born in Bethlehem, Pennsylvania, and lived there until 1798 when his family moved to Europe. He studied theology at the Moravian seminary in Niesky in present-day East Germany. In 1812 he was appointed Administrator of the Church Estates in North Carolina, and he and his bride started out for North America. They reached Salem, North Carolina, four months later, after spending fifty-eight days at sea. Once they were settled, Schweinitz combined church business trips with botanizing whenever he could. By the time he went to Europe in 1818, his list included 1,373 species of fungi, about 320 of which were described as new by him. In 1822 he moved to Pennsylvania. Altogether he described approximately 1,400 species of plants and fungi including many familiar mushrooms such as *Lactarius indigo* and *Cantharellus cinnabarinus*.

During the second quarter of the nineteenth century, two men became leaders in the study of southern fungi—Moses Ashely Curtis (1808–72) and William Henry Ravenel (1814–87). Curtis was a native of Stockbridge, Massachusetts. His first extensive contact with the South came when he was employed as a tutor in Wilmington, North Carolina, after he graduated from college. In 1835 he was ordained in the Episcopal Church and took up missionary work in western North Carolina. His headquarters were at Lincolnton, but he traveled widely in the area in pursuit of his duties and hobbies. On some of his trips he used a sulky with his portfolio for pressing plants under his seat. At various times he lived in Raleigh and Washington, North Carolina. In 1841 he settled in Hillsborough. It was his home for the rest of his life except for the period of 1847–56 when he served a pastorate in Society Hill, South Carolina. In addition to serving his church faithfully and energetically, Curtis led a second life as a botanist. His interest in fungi was evident by 1846 when he and a British clergyman, Miles Joseph Berkeley (1803–89), began a correspondence and specimen exchange. Although the two men never met, they coauthored numerous papers including several prepared by Berkeley that appeared after Curtis died. Curtis sent over 6,000 collections to Berkeley—his own as well as those of others in North America interested in fungi. Encouraged by Berkeley, Curtis began eating wild mushrooms and became an advocate of mycophagy. He believed that better knowledge of edible mushrooms in the South could have helped alleviate hunger during and after the Civil War. He prepared a manuscript on the edible mushrooms of North Carolina, but it was never published. He tested the edibility of at least 40 species that grew within two miles of his home, and, in a catalog of the plants of North Carolina, he indicated 111 species of edible fungi that occurred in the state.

Ravenel was born near St. Johns, South Carolina. Until 1853 he was a planter along the Santee River, then he moved to Aiken where he resided for the remainder of his life. During his years as a planter, he sometimes collected with a descendant of Thomas Walter. Ravenel's serious botanical work began in 1841. By 1846 he was interested in fungi and at about that time he and Curtis began corresponding. They became close friends and exchanged visits as well as specimens. By 1850 Ravenel reported that he had found 1,338 kinds of nonflowering plants in the region of his plantation. He prepared and issued the first sets (called exsiccati) of dried, identified fungi from North America. Ravenel, like Curtis, visited and examined Schweinitz's collections in Philadelphia and also corresponded with Berkeley. The Civil War left Ravenel impoverished and with a large family to support. He turned to botany as one source of income. As one means of raising money, he proposed to issue a set of exsiccati with another mycologist, Mordecai Cubitt Cooke (1825–1914) of England. Ravenel traveled to Georgia and Florida in the process of gathering material for this project. More than 50 plants, including fungi, were named for him. He became one of the best-known botanists in North America in his day and was respected both here and abroad.

Although Ravenel regarded the work he, Curtis, and others were doing as the work of pioneers, these two men knew more about the fungi of North America than any of their contemporaries. They corresponded with many fellow enthusiasts in this country and abroad and an active network of people interested in southern fungi developed around them. From their many correspondents, we have chosen to highlight two who made significant contributions, in different ways, to the study of southern fungi.

Thomas Minott Peters (1810–88) was a native of Tennessee, a lawyer, a politician, and eventually chief justice of the Alabama Supreme Court. His avocation was botany; his specialties were ferns, sedges, and fungi. Most of his collections were made in northwestern Alabama and were shared with appropriate specialists here and abroad. He represents the best attributes of amateur botany.

In contrast to Peters, Charles Horton Peck (1833–1917) was a botanist by profession. He was born in New York and, like Schweinitz, became interested in natural history as a child. From 1867 to 1915 he worked for the State of New York, becoming state botanist in 1883. Peck and Curtis corresponded during the latter part of Curtis's life, and when Curtis decided to sell duplicates from his personal collections, Peck bought a set. Peck's publications dealt with the plants and fungi of New York primarily; however, he published observations and descriptions of mushrooms sent to him from other areas also. In time he became the dean of American mycologists. Like Curtis, Peck was interested in mycophagy.

The economic and personal hardships of the Civil War turned men's thoughts from mycology to survival. Correspondence between Curtis, Peters, and Ravenel, and botanists in the North and in Europe were interrupted for years. Gradually in the latter part of the century a new set of faces emerged in southern mycology, but they usually turned to the North for assistance.

Mary Elizabeth Banning (1832–1901) spent much of her life in or near Baltimore, Maryland. She is the first woman we have encountered who published on southern mushrooms. Between 1877 and 1891 she wrote four papers on Maryland fungi. Some were quite casual and chatty, and in them she commented in a humorous manner on the attitudes of people toward mushrooms and mushroom hunting. Other papers included formal descriptions of new species. She produced a set of 175 plates of watercolor illustrations of mushrooms, and, in an accompanying manuscript, she described them and indicated which species were edible. Peck was her mycological mentor, and her paintings and manuscript were sent to him. He published descriptions of over a dozen new species which he credited to her.

George Francis Atkinson's interest in mushrooms was awakened during the time he spent in the South. Atkinson (1854–1918), a native of Michigan, completed his graduate work at Cornell University in 1885. From 1885 to 1892 he held appointments in the University of North Carolina, the University of South Carolina, and the Alabama Polytechnic Institute (now Auburn University). In 1888 and

1889 he spent the month of August in Blowing Rock, North Carolina, and while there found that: "The great prodigality of the fleshy fungi tempted the writer . . . to form the beginning of a closer acquaintance with the *Hymenomycetes* than had been gained from a general study of structure and relationships" (Atkinson and Schrenk 1893, p. 96). At first he turned to others for identifications of his mushrooms. As his interest in the subject grew, he devoted more and more time to mushrooms especially after he returned to Cornell as a member of the faculty in 1892.

The names of Lucien Marcus Underwood and Franklin Sumner Earle come up whenever one examines the history of mycology in Alabama. Underwood (1853–1907) was on the faculty of the Alabama Polytechnic Institute for the academic year of 1895–96. He collected fungi in several parts of Alabama, sent some to Peck who described several new species from this material, and published on others himself. Earle (1856–1929) joined the faculty in 1896 and remained there until 1901. Underwood and Earle prepared a report on Alabama fungi that was published in 1897. It was based primarily on their own collections and those of Atkinson and Peters, as well as reports in the literature. Underwood went on to a distinguished career at Columbia University. Earle joined the New York Botanical Garden and later became an expert in the culture of sugarcane. Some of the collections that the Earles made, particularly along the Gulf Coast in Mississippi, were described as new species by Murrill in later years. Mrs. Earle is among the first women to receive credit for an interest in southern mushrooms, although several wives must at least have tolerated their husbands' interest in fungi.

The era of the professional mycologist was beginning; amateurs, nevertheless, continued to play an important role. While most students of fungi were in communication with others of like interests, some worked in virtual isolation. One of these was Auguste Barthélemy Langlois (1832–1900), a Catholic priest who emigrated to North America from France in 1855. About 1857 he settled at Pointe-à-la-Hache, Louisiana, where he resided for approximately thirty years before moving to St. Martinville. He began to study the plants of Louisiana soon after his arrival, and sent specimens to France and probably eagerly awaited a reply. Apparently it never came, and he abandoned his studies. Twenty years later he heard of a botanist in New York and from him learned that there were many people in North America and Europe who shared his interests. Langlois returned to botany with renewed vigor and soon developed a large correspondence and specimen exchange. In 1885 he turned his attentions to fungi and before long worked with them almost exclusively. Like Curtis and Peck, he was a mycophagist and tried to interest others in the subject.

Three men began publishing on fungi in the early years of this century who deserve a place in our story: Henry Curtis Beardslee (1865–1948), William Alphonso Murrill (1869–1957), and William Chambers Coker (1872–1953). By this time Peck was solving Ravenel's dilemma by publishing notes on fungi in general and undertaking comparative accounts of various groups. These men adopted this approach also. Beardslee moved to Asheville, North

Carolina, from his native state of Ohio in 1901. He was interested in the mushrooms of western North Carolina and published several papers on them himself and in conjunction with Coker. After retiring from teaching, he spent his winters in Florida where he continued to study mushrooms.

Murrill was in many respects an idealist who found himself in an imperfect world. He was born near Lynchburg, Virginia, and lived in that state until he went to Cornell University for graduate studies in botany. At Cornell, he met and worked some with Atkinson. In 1900 Murrill received his doctorate and went to New York City where he taught high school biology until he moved on to a position at the New York Botanical Garden. He left there about twenty years later after experiencing a series of personal, financial, and health problems, and eventually settled in Gainesville, Florida. One summer, instead of going to Virginia as had been his custom, he stayed in Gainesville. Murrill described the result in his autobiography.

> Only then did he [Murrill] come to realize the richness of the endemic biota, and especially the vast opportunity in the fleshy fungi, which caused him to return to this group almost exclusively in later years. Many hundreds of new species were discovered and described in this difficult group. [1945, p. 148]

With the aid of many friends who brought him specimens and/or took him on collecting trips, over 600 new species were described by him from Florida between 1938 and 1955. Murrill was a man of many interests: the Girl Scouts, horticulture, music, and religion were just a few of them.

Coker was born in Hartsville, South Carolina. Like so many of the people we have encountered, his interest in natural history developed in his youth. After earning his doctorate in botany at the Johns Hopkins University he joined the faculty of the University of North Carolina at Chapel Hill, and resided there for the rest of his life. His first paper on fungi did not appear until 1907; mushrooms were only one of several aspects of botany that interested him. In addition to Beardslee, Alma Holland Beers and John N. Couch were important collaborators of his on studies of fleshy fungi. Coker was active in the development of the Highlands Biological Station in Highlands, North Carolina, and served it in many capacities over the years. He regularly spent his summers in Highlands and published on many new species from that region.

Gertrude Simmons Burlingham (1872–1952) was a native of New York, earned her doctorate at Columbia, and taught high school biology in Brooklyn. Whenever she could make time for it, she devoted herself to the study of the Russulaceae. Her mycological headquarters were at the New York Botanical Garden where she was an associate of Murrill's. Murrill's account of a collecting trip she took to the Pisgah Forest of North Carolina indicates she had a mischievous sense of humor. She and three others were collecting mushrooms when they were told by a young forester, who apparently wanted to startle the women, that rattlesnakes were good to eat. The botanists killed, skinned, and cooked a fat snake, and invited the forester to supper. He took one look at the long dish on the table and what was in it and "developed a sudden illness and had to leave for home immediately." One wonders what the women ate after they quit laughing. In her retirement years she spent the winters in Florida, where she and Beardslee prepared a paper on some local mushrooms.

The final student of southern mushrooms to be introduced here is Lexemuel Ray Hesler (1888–1977), better known as "Dean" or "Lex." He grew up in Indiana where his interests in natural history, athletics, and music became evident. He earned his doctorate in plant pathology at Cornell University. In 1919 he joined the faculty of the University of Tennessee where he remained throughout his life. He was an ardent collector and became a specialist in the fleshy fungi of the southern Appalachians. Along with Coker, he was active in developing the Highlands Biological Station, and he served as its vice-president for several years. He wrote numerous papers and several books, some of which were coauthored with Alexander H. Smith. Hesler met Peck and Atkinson and was acquainted with many mycologists in this country and around the world.

This brings us to the present, more or less, and current workers, both professional and amateur, who are too numerous to mention here. In spite of the contributions of past and present workers, the mycological wealth of the South still contains hidden riches in terms of undescribed species and extensions of known ranges of described species. The men and women who first "mined" this wealth found it exciting and stimulating, we hope you will also.

Key to Major Taxa
of Fungi Illustrated

1. Fruiting body consisting of a cap with gills on its lower surface; stalk present or absent 2
1. Not as above 3
   2. Gills with sharp edges even on button mushrooms, seldom interconnected, variously attached . . . . (p. 65) Agaricales
   2. Gills with blunt edges on button mushrooms, typically interconnected at maturity, decurrent . . . .
     . . . . (p. 59) Cantharelloid Aphyllophorales
3. Fruiting body with a cap bearing tubes (opening by pores) on its lower surface; stalk present or absent 4
3. Not as above 5
   4. Fruiting body consistently associated with wood, tough to woody at maturity; stalk present or absent; tube layer not easily and cleanly separable from cap . . . .
     . . . . (p. 45) Poroid Aphyllophorales
   4. Fruiting bodies of most species not consistently associated with wood, fleshy to soft at maturity; stalk typically present; tube layer typically easily and cleanly separable from cap . . .
     . . . . (p. 67) Boletaceae (Agaricales)
5. Fruiting body either with pendant spines or branched and corallike . . . 6
5. Not as above 7
   6. Fruiting body bearing spines, cap present or absent . . . .
     . . . . (p. 50) Spinose Aphyllophorales
   6. Fruiting body erect, branched, and corallike . . . .
     . . . . (p. 52) Clavarioid Aphyllophorales
7. Fruiting body of one of the following types: (1) resembling a gilled mushroom but with a layer of "pimples" instead of gills; (2) club-shaped and with a layer of "pimples" over the upper part; (3) cup-, saucer-, or funnel-shaped (and then arising from fallen branches or twigs) but not jellylike in consistency; or (4) consisting of a thimblelike, wrinkled, or pitted "head" and a distinct stalk, not foul smelling at maturity . . . . (p. 23) Ascomycotina
7. Not as above 8
   8. Fruiting body jellylike in consistency, ear- or shallowly cup-shaped, lobed, or taking shape around stems and debris . . . .
     . . . . (p. 41) Tremellales and Auriculariales
   8. Not as above 9
9. Fruiting body of one of the following types: (1) resembling a single funnel or a mass of funnels; (2) resembling a gilled mushroom but with a smooth hymenophore; or (3) resembling a cluster of erect, white to tan petals . . . . (p. 44) Aphyllophorales
9. Fruiting body of one of the following types: (1) resembling a small potato and developing at or just under the surface of the ground, interior neither powdery nor slimy at maturity; (2) resembling a small to large ball with the interior powdery at maturity; or (3) arising from an "egg" and having a slimy, malodorous spore mass at maturity . . . . (p. 241) Gasteromycetes

Ascomycotina

Ascomycotina

Members of the subdivision of the true fungi or Eumycota produce
the spores of their sexual state (the one where genetic recombination may occur) *in* specialized cells called asci in contrast to the
Basidiomycotina that form spores on specialized cells. This is a
large group that includes many economically important fungi including the yeasts used in brewing and baking. Of the six classes
sometimes recognized we discuss only a few examples of the Discomycetes in the order Pezizales and of the Pyrenomycetes in the
order Sphaeriales.

Key to Orders

1.   Some part of fruiting body covered with minute bumps or "pimples"
     from which white powder (spores) may be released; fruiting body
     club- or finger-shaped or resembling a distorted mushroom . . . . . .
     . . . . (p. 36) Sphaeriales
1.   No part of fruiting body covered with "pimples," spore-producing
     region smooth, wrinkled, or distinctly pitted; fruiting body variously
     shaped but not as above . . . . (p. 25) Pezizales

Pezizales

This is one of the two orders whose members are commonly called Discomycetes or cup fungi. The other order, the Helotiales, is not discussed here. The hymenium, in our examples, is exposed to the air by maturity and consists of asci and associated structures. When they are mature, the spores are forcibly shot out of the asci. Sometimes one can hear the hiss and see the miniature cloud of spores that results when many asci discharge simultaneously. Although these fungi are often called cup fungi, the fruiting body is not necessarily cup- or saucer-shaped.

For mycophagists, the Morchellaceae or morels, and the Helvellaceae or lorchels, are the most important families.

Key to Taxa

1. Fruiting body consisting of a distinct stalk and a specialized fertile portion or "head" that is thimblelike, wrinkled, lobed, or pitted 2
1. Fruiting body basically cup-, saucer-, or urn-shaped, with or without a stalk 3
   2. Head dull red to brownish red . . . . (p. 28) Helvellaceae
   2. Head tan to brown or black or a mixture of these, lacking red tints . . . . (p. 30) Morchellaceae
3. Fruiting in sandy areas, fruiting body more or less sunken in the sand; interior of cup brown . . . . 1. *Peziza ammophila*
3. Fruiting bodies associated with sticks and branches, not sunken into the sand; interior of cup red or blackish 4
   4. Upper portion of fruiting body red, shallow cup-shaped . . . . 2. *Sarcoscypha occidentalis*
   4. Fruiting body urn-shaped or funnel-shaped, often lacking a distinct stalk; gray to brownish black by maturity . . . . 3. *Urnula craterium*

Peziza ammophila Durieu & Mont.   **1**

Identification marks From above these fruiting bodies resemble brown holes in the sand with lobed edges, but when excavated the tapered base and spreading, lobed top are evident. The interior is dark brown; the exterior is so covered with sand its color is obscured. Sometimes a false stalk of hyphae and sand grains is present.

Edibility Too thin and too sandy to consider.

When and where This species was described from Algeria, is known to occur in Europe, and is shown here from South Padre Island, Texas. It is to be expected in coastal dunes, generally, but the details of its distribution remain to be learned. It fruits among sand dunes, especially in damp depressions, in the fall.

Microscopic features Spores 16–18 x 8–10 μm, smooth. Asci bluing at the tip in Melzer's reagent.

Observations Peziza is an ancient term that refers to a mushroom without a root or stalk; *ammophilus* means sand loving.

1 *Peziza ammophila*   Slightly under natural size

---

2   *Sarcoscypha occidentalis* (Schw.) Sacc.

Identification marks   These small fruiting bodies seldom exceed 1 cm in diameter. The shallow, disc-shaped cup with its bright red hymenium is borne on a whitish, distinct stalk.

Edibility   Of no consequence as an esculent.

When and where   Solitary or gregarious, arising from buried or partly buried twigs and small branches of hardwood trees; late spring through the summer; widely distributed and common in the hardwood forests east of the Great Plains. Schweinitz described it from specimens collected in Ohio.

Microscopic features   Spores 18−20 × 10−12 μm, smooth. Asci operculate, not turning blue in iodine.

2 *Sarcoscypha occidentalis*   About natural size

26

Observations The scarlet cup, *S. coccinea*, is common in early spring in much the same habitats; it has larger cups (2–4 cm broad) and little or no stalk. *Sarcoscypha* means fleshy cup; the specific epithet means west or western—in Schweinitz's time Ohio was "western."

Urnula craterium (Schw. : Fr.) Fr.   3
(Black Tulip)

Identification marks These tough fruiting bodies resemble a hollow club at first (pl. 3a), but gradually open up (pl. 3b). At maturity they are vase-shaped or resemble a stemmed glass. The hymenium (lining of the cup) is dark brown to blackish, the exterior usually some shade of gray to nearly black in age.

Edibility Not recommended; it is tough.

When and where Solitary to clustered, often attached to dead wood, especially branches of hardwood trees that are partly buried; common but inconspicuous. It fruits in March or earlier in the deep South to May or early June in the North and is widely distributed in eastern North America. Schweinitz collected the type in North Carolina.

Microscopic features Spores 24–36 × 10–15 μm, smooth. Asci operculate, not turning blue in iodine.

Observations Like *Craterellus fallax* these fruiting bodies are often mistaken for black holes in the ground. In *Urnula* the spores are formed on the interior of the fruiting body; in *Craterellus* on the exterior. *Urnula* means a little urn; the specific epithet refers to the shape, that of a crater, a vessel used in antiquity in which wine and water were mixed.

3a   *Urnula craterium*, young specimens   Natural size

3b *Urnula craterium*, mature specimens   One-half natural size

Helvellaceae

Like the members of the Morchellaceae, the fruiting bodies in this family are often large, have a stalk and a distinct head (on the surface of which spores are produced), and asci that do not turn blue in iodine. The Helvellaceae differ in that the spores contain distinct oil drops, contain only four nuclei, and are ornamented in some species. Of more use to mushroom hunters are the following characters: the hymenium is white, tan, gray, brown, black, or dull ochraceous to deep brownish red; and the head typically is neither pitted nor thimble-shaped. In the species with a pitted head, the ribs are fertile and similar in color to the pits. In North America there are several genera in the family and about seventy species. The two species of *Gyromitra* illustrated here may be encountered in the spring by turkey hunters and morel hunters. Some species in this genus have caused severe poisonings. They can be recognized by the fact that the stalk is typically hollow by maturity and lacks internal folds and the head is rusty orange to brownish red. They do not seem to be common in the South. The name *Gyromitra* is derived from words meaning round and a bishop's miter.

Key to Species

1. Head formed into 2–3 lobes which are flattened and resemble wings or the ears of an elephant .... 4. *Gyromitra fastigiata*
1. Head pitted and wrinkled, more or less rounded in outline rather than lobed .... 5. *G. caroliniana*

28

Gyromitra fastigiata (Krombholz) Rehm   **4**
(Elephant Ears)

Identification marks   The lobes of the brownish red fertile portion or head are flattened and typically upraised at the tips like the horn of a saddle; otherwise the surface is somewhat wrinkled but not pitted. The stout massive stalk is whitish to chalky white and lacks pronounced ridges. Large specimens may weigh over a half pound.

Edibility   Early reports on this species in North America stated it was tender and well flavored; it was sold in markets in St. Louis around the turn of the century. In recent years its reputation has become somewhat tarnished. We do know several people who eat and enjoy it but we recommend caution.

When and where   Scattered to gregarious in rich, often low, hardwood forests; spring, about the time the first morels appear. In North America its apparent range is from the eastern edge of the Great Plains to the Atlantic Ocean excluding the coastal plain and extending only into the region of the Great Lakes on the north.

Microscopic features   Spores $28-30 \times 13.5-15\mu m$, ornamented with a distinct reticulum and one to several short apiculi at each end.

Observations   Older books may discuss this species under *G. brunnea* or *Helvella underwoodii*. *Fastigiatus* means with clustered erect branches producing a narrow elongated habit.

4   *Gyromitra fastigiata*   Two-thirds natural size

5 *Gyromitra caroliniana*   One-fourth natural size

5   *Gyromitra caroliniana* (Bosc : Fr.) Fr.
   (Brown Bonnets)

Identification marks   Numerous folds and wrinkles combine to form the characteristic wrinkles and depressions of the rounded fertile portion or head. It is not lobed as in *G. fastigiata*. Unlike the true morels (genus *Morchella*), the ridges and pits are similar in color and texture. The massive stalk is complex internally and has rounded ribs on the exterior.

Edibility   We have several reports that it is edible but do not recommend it unreservedly. Bosc was of the opinion it had neither odor nor flavor to recommend it.

When and where   Early March in the deep South into May along the southern edge of the Great Lakes; scattered to gregarious under hardwoods, often in damp areas; widely distributed in eastern North America outside the coastal plain and high mountains. It was described from "la haute Caroline" and is often abundant along the Mississippi River.

Microscopic features   Spores 25–30 × 12–14µm, ornamented with a distinct reticulum that forms one to several short apiculi at each end.

Observations   As expected, the specific epithet refers to the Carolinas where Bosc discovered it.

Morchellaceae

For mycophagists this is the most important group in the Ascomycotina: it is the family to which the morels belong. Morels, also called sponge mushrooms, generally fruit in the spring about the time apple trees bloom. In the South they are most abundant outside Florida and the coastal plain.

The species sought for table use have medium to large fruiting bodies with a distinct stalk and a head that is either thimblelike or resembles a sponge and has a pitted surface. The hymenium covers the outer surface of the head except for the ribs in some species. Technical characters of this family include spores with homogeneous contents, many nuclei, and smooth walls; and asci that do not stain blue in iodine. Except as noted, the asci contain eight spores. It is easy to recognize a morel in the field without checking the technical characters, but deciding on what name to use for the ones you find is not as easy. There are estimated to be five to ten "good" species of *Morchella* (true morels) in North America, but there is little agreement as to what names to use for them. The name *Morchella* is derived from an old German name for these fungi; *Verpa* means a rod.

Key to Species

1. Head attached only at apex of stalk, sides of head hang free . . . .
 . . . . 6. *Verpa conica*
1. Either head not hanging free of stalk at all or only half to two-thirds
 of its length hanging free 2
  2. Lower half to two-thirds of head hanging free of the stalk . . . .
 . . . . 7. *Morchella semilibera*
  2. At most only the margin of the head free from the stalk 3
3. Ribs black by maturity . . . . 8. *Morchella elata*
3. Ribs white to ivory when young and unchanging or becoming dull
 yellowish brown by maturity 4
  4. Pits predominantly round to irregular in outline but not gener-
 ally 2–3 times longer than broad . . . . 9. *Morchella crassipes*
  4. Pits predominantly 2–3 times longer than broad . . . .
 . . . . 10. *Morchella deliciosa*

Verpa conica (Müller : Fr.) Schwartz   **6**

Identification marks The fruiting body resembles a large thimble on a small finger; the head is attached to the fragile stalk at its apex. The sides of the head hang free of the stalk and are never strongly ribbed. The stalk is hollow.

Edibility Edible but mediocre in the opinion of many who try it.

When and where Solitary to gregarious, often under hardwoods or old apple or cherry trees; early spring about the time the first morels appear or a little earlier. It is widely distributed in North America, but we suspect it is rare to absent from the coastal plain and Florida.

Microscopic features Spores 20–26 × 12–15µm.

Observations   Few specimens are ever truly conic in spite of the specific epithet. *V. bohemica*, the early morel, has distinct ridges on the head, 2-spored asci, and large spores (60–80 × 15–18µm). It has caused a temporary loss of coordination in some who have eaten it.

6   *Verpa conica*   Slightly under natural size

---

7   *Morchella semilibera* de Candolle : Fr.
(Half-free Morel)

Identification marks   As suggested by the common and scientific names, part (the lower half to two-thirds) of the head hangs free of the stalk. At first and when covered by leaves, the ribs are grayish tan but they darken to black by maturity. Young specimens have a large head in relation to the stalk; in old ones the proportions are usually reversed.

Edibility   Edible and good; however if you are sensitive to the early morel (see observations) be especially careful when gathering specimens of *M. semilibera* for table use.

When and where   Scattered under hardwoods, old apple trees, or other fruit trees; spring, about the middle of the morel season; widely distributed but to be expected primarily in the northern and montane parts of the South.

Microscopic features   Spores 20–30 × 14–18µm.

Observations   In *Verpa bohemica*, the early morel, the head is attached to the stalk at its apex, the sides hang free and are more wrinkled than pitted; furthermore, the asci are 2-spored rather than 8-spored. The early morel may cause a temporary loss of coordination when eaten in quantity.

7 *Morchella semilibera*   About one-half natural size

Morchella elata Fr. : Fr.   8
(Black Morel)

Identification marks In young specimens the ridges appear to be made of fine gray velvet. As they age the ridges become blacker, thinner, and duller. The pits are some shade of brownish gray at all stages and lighter than the ribs at maturity. The fruiting bodies are hollow.

8 *Morchella elata*   One-half natural size

Edibility Fresh young specimens are excellent esculents; in age they become strongly and somewhat unpleasantly flavored. Some people experience mild to moderate digestive upsets after eating black morels in quantity, especially if alcohol is also consumed.

When and where Under conifers, apple trees, in mixed woods, or possibly under hardwoods in the spring. This black morel is known from the Pacific Northwest and is shown here from the Piedmont of North Carolina, but the details of its distribution remain to be determined.

Microscopic features Spores 24–27 × 13–15μm.

Observations *Elatus* means tall. A second species of black morel also occurs in North America. It has smaller spores (20–24 × 12–15 μm) and often has smaller fruiting bodies. This species is often called *M. angusticeps* but some authors lump both species into one and call it *M. conica*. Mycophagists seldom distinguish between the two species. We have no data on their relative merits as esculents. Of all the morels, the group of black morels is responsible for more gastrointestinal upsets than any other.

---

9 *Morchella crassipes* (Ventenat : Fr.) Pers. (Thick-footed Morel)

Identification marks Large size, as compared to other morels with pale ridges, is the outstanding feature of this species: fruiting bodies are often 15–20 (30) cm tall at maturity. The head is basically columnar to slightly tapered at first, and the pits are more or less rounded rather than elongate as in *M. deliciosa*. The pale creamy tan stalks appear to be gathered or pleated at the base.

Edibility Excellent, one of the most popular edible mushrooms in North America.

9 *Morchella crassipes*　　　　　　　　　　About one-third natural size

When and where Fruiting begins by late March along the lower Mississippi River and spreads northward, reaching the Canadian border in early or mid-May. Large fruitings are sometimes found around dead elms, under old apple trees, and in rich hardwood forests especially in floodplains. In the South it is widely distributed outside the coastal plain.

Microscopic features Spores 21–26 × 13–16 μm.

Observations The distinctions between this species and *M. esculenta* are those of degree; there is a distinct possibility that only one species is involved. Anyone, however, who finds a patch of these large, delicious morels is likely to agree they are something special. The specific epithet means thick footed.

Morchella deliciosa Fr. : Fr.   10

Identification marks This is often about the last species of morel to fruit in the spring. Its appearance marks the end of the morel season. The slender, short fruiting bodies have prominent, widely spaced vertical ribs and less prominent cross veins on the head that divide it into large pits. At first the pits are grayish tan to tan and the ribs lighter tan to cream color but by maturity both are usually dull yellowish tan.

Edibility Edible and good.

When and where Under old apple trees and in hardwood forests from March in the South into May in the North; widely distributed in eastern North America outside the coastal plain.

Microscopic features Spores 21–24 × 12–15 μm.

Observations Some authorities have called this species *M. conica*, but in Europe, the "home" of both species, that name is generally used for a species with ribs that blacken in age. Since none of the members of this genus is known to be dangerously poisonous, mycophagists can eat and enjoy where mycologists fret and ponder.

10 *Morchella deliciosa*   About natural size

Sphaeriales

In this order the asci are borne within miniature flask-shaped structures called perithecia that open to the air by a small pore called an ostiole. The perithecia may or may not be sunken in a mass of fungous tissue. When they are sunken they often resemble tiny pimples. Many of the species in this order are parasitic on other fungi, plants, or animals and are of considerable economic importance.

Key to Species

1. Fruiting body upright, finger- or club-shaped and attached to a buried, dead larva of a beetle (excavate specimens carefully to find the beetle larva) . . . . 11. *Cordyceps melolonthae*
1. Fruiting body duplex, consisting of a distorted gilled mushroom and the hyphae of the parasite; at maturity the parasite forms perithecia where gills would normally be formed . . . . . 12. *Hypomyces lactifluorum*

---

11   *Cordyceps melolonthae* (Tul.) Sacc.

Identification marks   This fungus obtains its nourishment at the expense of the larval stage of a June bug (the white grub that often causes trouble in lawns). The fleshy, fingerlike, dull yellow fruiting body develops a pimpled area near but not over the apex. Each pimple is a flask-shaped structure called a perithecium within which many asci are formed. The asci, of course, produce the spores—some of which presumably infect healthy white grubs.
Edibility   We have no data on it.

11   *Cordyceps melolonthae*   About natural size

When and where One "finger" is produced per grub. Only the finger projects above the ground; you must dig to find the grub. This species is widely distributed in eastern North America and fruits from June in the North on into the fall and winter in the South.

Microscopic features Spores threadlike, soon breaking up into segments (4) 6−8 (10) × 1.5−2.5 µm. Asci inoperculate.

Observations More species of Cordyceps have been reported from the South than from any other part of North America. Many species parasitize insects, but a few attack Ascomycetes such as Elaphomyces, which produce their fruiting bodies in the ground like a truffle. Cordyceps means club head; the specific epithet is the name of a genus of beetles whose members may be attacked by this fungus.

Hypomyces lactifluorum (Schw. : Fr.) Fr. 12

Identification marks These curious objects are gilled mushrooms parasitized by a member of the Ascomycotina. Small orange "pimples" on the area where one would expect to find gills mark the location of the perithecia produced by the parasite. In old specimens dots of whitish powder (actually spores) are often visible at the opening of each perithecium. In age and when touched with potassium hydroxide or ammonium hydroxide, the orange areas become magenta.

Edibility Generally rated edible but not recommended because of the difficulty of identifying the host.

When and where Common in late summer to early winter on members of the Russulaceae; widespread in North America, and common east of the Great Plains. Schweinitz collected it in North Carolina.

Microscopic features Spores 31−42 × 7−8 µm, 2-celled, roughened, pointed at each end.

Observations The parasite slows the process of decay in the host with the result that infected fruiting bodies last longer than uninfected ones. Hypomyces means below or under a fungus; the specific epithet refers to the milklike latex produced by species of Lactarius, a common host of this parasite.

12 Hypomyces lactifluorum   Two-thirds natural size

Basidiomycotina

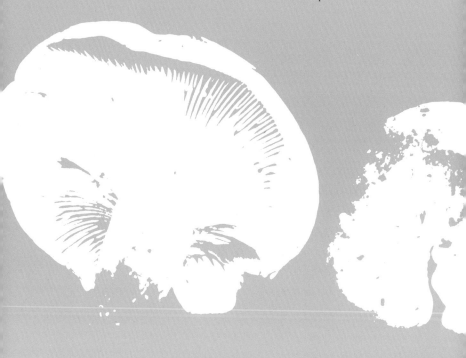

Basidiomycotina

Members of this subdivision of the true fungi or Eumycota produce the spores of their sexual stage (the one where genetic recombination may occur) *on* specialized cells called basidia in contrast to the Ascomycotina that form spores within asci. Three classes within the Basidiomycotina may be recognized: the Teliomycetes that are all parasitic on plants and are not discussed here, the Hymenomycetes in which the basidia are one- or more-celled and the spores are typically discharged at maturity, and the Gasteromycetes in which the basidia are one-celled and the spores are not forcibly discharged at maturity. Most of the fungi we discuss belong to one of the classes of the Hymenomycetes: the Tremellales, Auriculariales, Dacrymycetales, Aphyllophorales, or Agaricales.

Tremellales and Auriculariales

These two orders together with the Dacrymycetales constitute the group of fungi commonly called the jelly fungi because of the jellylike consistency of their fruiting bodies. In contrast to the members of the Aphyllophorales and Agaricales, in which each basidium consists of a single cell, basidia in the jelly fungi are divided into four cells by maturity. The shape of the basidium and the orientation of the septa are characters used in defining the orders of jelly fungi, but are often difficult to demonstrate.

Aside from a few edible species, this group has little economic importance. *Tremella* means jellylike or trembling; *Auricularia* means ear-shaped.

Key to species

1. Fruiting body reddish brown to brown, somewhat ear- to cup-shaped .... 13. *Auricularia auricula*
1. Fruiting body translucent, white, ivory, or pale tan 2
    2. Fruiting body translucent to white, with a single point of attachment .... 14. *Tremella fuciformis*
    2. Fruiting body more or less opaque, white to dull ivory; formed around pieces of debris, plant stems, etc. ....
    .... 15. *Tremella concrescens*

Auricularia auricula (L. : Fr.) Underwood   **13**
(Wood Ear)

Identification marks These tough, gelatinous to rubbery fruiting bodies are irregularly ear-shaped and have short ribs near the point of attachment. The spores are formed on the outer surface of the dark brown fruiting bodies.

Edibility Edible, particularly popular in the Orient and in oriental-style cooking. We soak dried specimens in warm water until they are soft, slice them, and use them in casseroles, etc., where they add interesting texture.

When and where Singly or in clusters on dead wood, especially hardwood logs with the bark still in place; fall and early winter; widely distributed and common in many parts of the world.

Microscopic features Spores 11−14 × 5−7 µm, smooth, hyaline.

13 *Auricularia auricula*   Three-fourths natural size

Observations   These fruiting bodies dry and rehydrate well. Large quantities of them are imported from the Orient although the species is common in parts of North America. There is some debate over the correct placement of this species, *Hirneola* being the alternate genus. *Auricula* means ear.

14   *Tremella fuciformis* Berk.
(Silver Ear)

Identification marks   This ghostly white fungus is jellylike and translucent when fresh but thin, pale tan, rigid, and almost lacy when dry. The fruiting bodies are distinctly and complexly lobed and are seldom more than 7 cm long and 4 cm high.

Edibility   It is considered edible in the Orient, we have no data on North American collections.

When and where   Solitary to scattered on wood of hardwoods; summer and fall during wet weather; widely distributed in the warmer parts of the world including the southeastern United States.

Microscopic features   Spores 6–11 × 5–7 (8.5) μm; smaller conidiospores also present.

Observations   This species was mentioned in Chinese pharmacology texts and was used both as a drug and a food. It is now being cultivated in several countries and is imported to this country from Taiwan. Another common name for it is white jelly mushroom. The specific epithet apparently means shaped like seaweed or resembling a foliose lichen.

14   *Tremella fuciformis*   Natural size

15   *Tremella concrescens* (Schw. : Fr.) Burt

Identification marks   These irregular, tough to gelatinous fruiting bodies seem to develop in a random manner. The fruiting bodies are whitish to pale tan, hollow, and use stems of plants and bits of detritus as a means of support—in our illustration a live poison ivy plant! The poison ivy is more dangerous to many hunters than the mushrooms themselves and lends emphasis to the admonition to "look before you pick!"

Edibility   Who cares? Not recommended, especially when it fruits on poison ivy.

When and where   Late spring into fall in woods, shaded lawns and gardens, and similar places. It is widely distributed in the central and eastern United States and reported from South America. Schweinitz collected it in North Carolina.

Microscopic features   Spores 9−11 × 5−7 μm. Clamp connections absent.

Observations   The specific epithet means curdled or congealed.

15   *Tremella concrescens*   Slightly under natural size

Aphyllophorales

Basically any fleshy fungus that bears spores on basidia and forcibly discharges them, is not gelatinous, and is neither a bolete nor a gilled mushroom belongs to the Aphyllophorales. The form of the hymenophore (the spore-producing region of the fruiting body) may be smooth, wrinkled, foldlike, spinose, tubular (but if so the fruiting bodies are tough and not readily decaying), or somewhat gill-like. This is a diverse order in terms of fruiting body morphology.

We do not divide this order into families because the technical characters used to separate most of them are beyond the scope of a field guide. Instead, the order is divided into groups based on similarities in gross morphology. Some species we include do not fit into the main groups and are treated individually in the following key.

Key to Taxa

1. Fruiting body whitish to tan and in the form of a cluster of flattened upright segments resembling in mass a head of leaf lettuce or a fully opened rose 2
1. Not as above 3
    2. Both surfaces of the segments smooth at maturity ....
        .... 16. *Sparassis spathulata*
    2. Outer surface of segments with low ridges that form a reticulum, distinct tubes, or irregular teeth ....
        .... see 17. *Hydnopolyporus palmatus*
3. Hymenophore either of separate tubes resembling miniature pipes or poroid with united tubes; fruiting bodies rather firm to fibrous at maturity .... (p. 45) Poroid Aphyllophorales
3. Hymenophore not as above 4
    4. Hymenophore in the form of pendant spines ....
        .... (p. 50) Spinose Aphyllophorales
    4. Hymenophore smooth, wrinkled, or somewhat gill-like 5
5. Fruiting body branched and corallike ....
    .... (p. 52) Clavarioid Aphyllophorales
5. Fruiting body funnellike and hollow or resembling a gilled mushroom, i.e., with a distinct cap and stalk ....
    .... (p. 59) Cantharelloid Aphyllophorales

---

16   *Sparassis spathulata* (Schw. : Fr.) Fr.

Identification marks   If you find a fruiting body that resembles a head of leaf lettuce or a large rose you probably have either this species or *Hydnopolyporus palmatus*. In the *Sparassis*, the segments are thin, firm, erect, and persistently smooth on both surfaces. They are pale ivory to buff and banded with zones of deeper color.

Edibility   Edible and good when young and tender.

When and where   Solitary to scattered, often at the base of stumps of living or dead trees; summer and fall; widely distributed in eastern North America but most abundant in the South.

16   *Sparassis spathulata*   Three-fourths natural size

Microscopic features   Spores 6–7.5 × 4.5–5.5 µm, smooth.

Observations   *S. crispa*, the common species in western North America, has a prominent rootlike base and lax, petallike segments. Both are edible and have been called cauliflower mushrooms. *S. spathulata* and *Hydnopolyporus palmatus* are easily confused unless one takes time to note that the fruiting bodies of the latter are whiter and have short to prominent pores or teeth on the outer surface of each lobe or segment. *Sparassis* is derived from the Greek for "I tear to pieces"; *spathulatus* means spathulate, i.e., with a broad flattened apex tapered down to a narrower stalk. Some authors prefer to call this species *S. herbstii*.

Poroid Aphyllophorales

Fruiting bodies that are tough to woody or fibrous at maturity and have a poroid hymenophore are characteristic of this group. Fruiting bodies may last a few weeks or even be perennial. The tube layer is not readily separable from the cap as it is in the Boletaceae. It is often difficult to find spores on polypores because many are slow to mature and spore production may be interrupted without significant change in the appearance of the fruiting bodies.

   Most members of this group decay wood; some cause serious heart rots of living trees; others decay dead trees, logs, and stumps. Only a few polypores have fruiting bodies that are soft enough when young to be edible. This is a large and diverse group.

Key to Species

1.   Fruiting body resembling a cluster of erect to ascending whitish to pale tan, firm, thin petals with pores, spines, or a low reticulum on the outer surface .... 17. *Hydnopolyporus palmatus*
1.   Not as above 2

---

17   *Hydnopolyporus palmatus* (Hook.) O. Fidalgo

Identification marks   Numerous thin, spathulate or lobed, erect, white to pale cream segments grow in rosettelike clusters that may be up to 10 cm broad. At most only a short stalk is present on each segment. The outer (lower) surface, where the hymenium is located, varies from almost smooth to reticulate, distinctly poroid, or toothed with flat teeth.

Edibility   Dan Guravich reports it is edible but inclined to be chewy. It is eaten by some Brazilian Indians and sometimes sold in markets in Mexico.

When and where   Early summer through fall on stumps or arising from buried wood of broad-leaved trees. It was described from Columbia, South America, and extends into the United States around the Gulf of Mexico. It also occurs in the West Indies.

17   *Hydnopolyporus palmatus*   About one-third natural size

Microscopic features Spores 3.5–5 × 2.5–3.5 μm, smooth, hyaline, inamyloid.

Observations Especially when the pores or teeth are poorly developed, this species might be mistaken for *Sparassis spathulata* and eaten in its stead. The genus name refers to the toothed and poroid nature of the hymenophore; *palmatus* means lobed or divided like a hand with fingers.

18 *Meripilus giganteus*   One-fourth natural size

Meripilus giganteus (Pers. : Fr.) Karst.   **18**
(Giant Polypore)

Identification marks Dull gray large clusters of petallike, fleshy, shelving caps, each with little or no stalk, are arranged in compact rosettes. A single fruiting body may have from two to fifteen or more caps. In age, when injured, and when dried, dark brown to blackish stains develop on the caps and pores.

Edibility Young, tender, thoroughly cooked specimens are edible for most people; however, this species is the apparent cause of occasional gastrointestinal upsets so try it in moderation—if at all.

When and where On the ground, usually near stumps or trees, especially oak and beech; summer and fall; widely distributed east of the Great Plains and not rare in the South.

Microscopic features Spores (5) 6–7 × (3.5) 4.5–6 μm, smooth, hyaline, inamyloid.

Observations We suspect this species has been misidentified as unusual material of hen of the woods (*Grifola frondosa*) and eaten more often than reports indicate. Hen of the woods has small, thin-fleshed caps and neither stains nor dries black. Fortunately, it is also edible when young and tender. *Meripilus* means many caps, *giganteus* means very large or gigantic.

19 *Ganoderma curtisii* (Berk.) Murr.

Identification marks   In young specimens the stalk and cap appear to be coated with a dull ochraceous to dull red varnish. It often wears away in age. The context is soft but not readily chewed; it is white near the surface and light brown near the tubes. The pores are minute; their mouths quickly stain brown when bruised.

Edibility   Too tough to chew, better used as a table ornament.

When and where   Arising from buried wood or on stumps or trunks of hardwood, particularly oak, magnolia, sweet gum, maple, and locust. It is widely distributed in eastern North America and especially abundant in the Southeast. It can be found throughout the year.

Microscopic features   Spores 9–11 × 5–7 μm, minutely echinate, light brown in KOH.

Observations   M. A. Curtis is commemorated by the specific epithet. The context of these fungi darkens to black or nearly black when touched with KOH. Dried specimens can be attractive decorations by themselves or in dried arrangements. If they get "buggy" they can be fumigated by placing them and a strip of insecticide-impregnated paper in a plastic bag for a few days.

19 *Ganoderma curtisii*   About one-half natural size

20 *Pycnoporus sanguineus* (L. : Fr.) Murr.

Identification marks   These brilliant orange-red bracket fungi are too tough to eat but easy to admire. The color persists in dried specimens and they are used at times to lend color to dried arrangements. The surface of the cap has been described as appearing as if it had been seared with a hot iron. When touched with KOH, it turns magenta or black.

Edibility   Inedible.

When and where   On wood of hardwoods, especially oak, beech, sycamore, tupelo, and sweet gum; common along the Gulf of Mexico and reported as far north as New York and Iowa. Fruiting bodies may be found throughout the year but may not always have spores.

Microscopic features   Spores 4–5 × 2–3 µm, often absent or hard to find.

Observations   The genus name means close- or dense-pored; *sanguineus* means blood red. *P. cinnabarinus* is similar in color but has softer, thicker fruiting bodies that are wrinkled to tomentose on the upper surface. It is a northern species although the ranges of the two overlap.

20   *Pycnoporus sanguineus*   About natural size

Laetiporus sulphureus (Bull. : Fr.) Bond. & Sing.   **21**
(Sulphur Shelf)

Identification marks   When fruiting on the side of a tree or log these fleshy fruiting bodies resemble a series of shelves, but when arising from buried wood or at the base of a tree, they often form giant rosettes. Each shelf or lobe is bright orange to salmon orange on top but the margin and the underside are lemon to sulphur yellow (or white in var. *semialbinus*).

Edibility   Thoroughly cooked, tender, young specimens are edible for most people and are popular in some regions. This species, however, does cause gastrointestinal upsets (especially if eaten raw). Specimens growing on conifers or *Eucalyptus* trees tend to have stronger flavors and are apparently more likely to cause illness.

21   *Laetiporus sulphureus*   About two-thirds natural size

When and where   On dead wood or on living but dying trees, especially oak and willow in eastern North America but found on many kinds of trees; summer to early winter; widely distributed in North America.

Microscopic features   Spores 5–7 × 3–4.5 µm, smooth, hyaline, inamyloid.

Observations   As specimens age, the colors fade until old ones may be nearly white. The species has long been considered "safe" for mycophagists; evidence to the contrary, however, is accumulating. *Laetiporus* means bright pores; *sulphureus* means sulphur yellow. Another common name for it is chicken mushroom.

Spinose Aphyllophorales

Members of this group are united in having a spinose hymenophore, i.e., the hymenium covers pèndant spines. The fruiting body may consist of a distinct cap and stalk with the spines on the underside of the cap, or the spines may be borne on a lump of tissue or a branching framework. There are many species in this group in the South that have tough to woody fruiting bodies and/or strong, unpleasant flavors. These should not be eaten. We illustrate two species however, that are edible and considered good by most mushroom hunters.

In classical times *Hydnum* as a name was used for some truffles but for many years it has been applied to a group of spine-bearing fungi; *Hericium* refers to hedgehogs.

Key to Species

1.   Fruiting body consisting of a cap with pendant spines on the lower
     surface and a stalk .... 22. *Hydnum repandum*
1.   Fruiting body consisting of numerous spines hanging from a fleshy
     mass of tissue .... 23. *Hericium erinaceus*

Identification marks These stout, fleshy fruiting bodies have slender, fragile spines on the lower side of the cap. The cap surface is smooth at first but becomes variously cracked (rimose) as it ages. In var. *repandum* the color is pale reddish cinnamon to pale orange tan; in var. *album* all parts are white. In both varieties injured areas quickly stain yellow and the taste is mild to peppery. The spores are white in deposits.

Edibility Mild specimens are excellent for eating.

When and where In woods generally, especially where oak and/or blueberries occur, in eastern North America but widely distributed; summer and fall; often abundant.

Microscopic features Spores $6-8 \times 5.5-7$ µm, inamyloid.

Observations *Repandus* means having a spreading, wavy margin. Two species with white fruiting bodies may also be encountered in the South: *H. albidum* which stains orange when bruised, has acrid flesh, and spores $4-5.5 \times 3.5-4$ µm; and *H. albomagnum* which does not stain, has a mild taste, and spores $5.5-7.5 \times 3.5-4.5$ µm. None of these species are known to be poisonous. Many mushroom hunters know these as species of *Dentinum*, the change in name to *Hydnum* reflects recent changes in the nomenclature of this group.

22 *Hydnum repandum* var. *album*   About two-thirds natural size

Hericium erinaceus (Bull. : Fr.) Pers.   **23**
(Hedgehog Mushroom)

Identification marks The first impression one may have of these fruiting bodies is of a patch of coarse polar bear fur tacked onto a tree or log. Pendant spines up to 4 cm long are borne on a fleshy mass called a tubercle.

When young the spines and tubercle are white; they often become dull yellow or brownish in age.

Edibility   Edible and good when young and tender.

When and where   Usually solitary; arising from trees, stumps, and logs of broad-leaved trees, particularly oak, beech, maple, and sycamore; summer and fall to early winter. It is widely distributed in North America but most abundant in the South.

Microscopic characters   Spores 5.5−7 × 4.5−5.5 µm, finely roughened to smooth, amyloid, white in deposit.

Observations   *Erinaceus* means hedgehog. *H. coralloides* has coarse spines borne at the tips of branches that form a framework. It also fruits on the wood of broad-leaved trees and is more likely to be found from North Carolina and Tennessee northward.

23   *Hericium erinaceus*   About one-half natural size

Clavarioid Aphyllophorales

Fruiting bodies that are erect simple clubs, or branched and coral-like, in which the surface of the club or branches is smooth to slightly wrinkled are characteristic of this group. The spores are produced on the outer surface of the fleshy, often fragile, fruiting bodies. Although some species are edible and good, many have not been tested. Mycophagists will find that making identifications in this group is not a simple task; the characters used to define the genera are mostly technical and outside the scope of a field guide. We include several species as examples of diversity in the coral fungi. *Clavaria* refers to a club, *Clavulina* to a small club; *Ramaria* means branched, and *Ramariopsis* means *Ramaria*-like.

Key to Species

1. Fruiting body white to light gray, dark gray, or lavender to violet overall and not staining green with $FeSo_4$ ...., ... 2
1. Fruiting body brown, tan, orange, rose, dull yellow, or lavender in part and staining green with $FeSo_4$ 4
    2. Fruiting body lavender to violet .... 24. *Clavaria zollingeri*
    2. Fruiting body white or gray 3
3. Fruiting body white, usually staining apricot to salmon where injured .... 25. *Ramariopsis kunzei*
3. Fruiting body dingy white to light or dark gray, not staining as above .... 26. *Clavulina cinerea*
    4. Base of fruiting body compound, seemingly composed of several stalks pressed together .... 27. *Ramaria conjunctipes*
    4. Not as above 5
5. Base of fruiting body thin; upper branches light to deep reddish brown .... 28. *Ramaria murrillii*
5. Base of fruiting body fleshy, thick; upper branches differently colored than above 6
    6. Base of fruiting body and branches dull lavender to dingy violet when young, in age tan to light brown .... 29. *Ramaria fennica*
    6. Base whitish to pale yellow or salmon orange; branches salmon orange to coral pink when young, fading to yellow in age .... 30. *Ramaria subbotrytis*

Clavaria zollingeri Léveillé   24

Identification marks   These amethyst to violet fruiting bodies are fragile and sparingly to much branched. Frequently the branching pattern is a dichotomous one—dividing into two at each fork. The fruiting bodies are seldom large. Neither the odor nor the taste are pronounced.

Edibility   Not reported.

24   *Clavaria zollingeri*   About natural size

When and where   On the ground in deciduous or mixed woods; late summer and fall; widely distributed in eastern North America outside the coastal plain but seldom abundant.

Microscopic features   Spores 6–7.5 ×3–4 µm, smooth, white in deposits. Basidia 4-spored. Clamp connections absent.

Observations   Two species of *Clavulina*, *C. amethystinoides* and *C. amethystina*, have lilac to purple fruiting bodies similar to those of *Clavaria zollingeri*; however, in the species of *Clavulina* basidia are 2-spored and clamp connections are present. In *Clavulina amethystinoides* branching is often palmlike to toothed, but the three species may be difficult to distinguish in the field. Heinrich Zollinger (1818–59) is commemorated in the specific epithet; he was born in Switzerland and died in Java. The coral fungi were of special interest to him.

---

25   *Ramariopsis kunzei* (Fr. : Fr.) Donk

Identification marks   These much-branched fruiting bodies are chalky white at first but may become pinkish to pale salmon where injured or in age. The branches are less likely to break from rough handling than those of some coral fungi.

Edibility   Reported to be edible.

When and where   Widely distributed and common in the forested areas of North America; known from the Gulf Coast north into Canada. It fruits in the summer and fall.

Microscopic features   Spores (3) 4–6 × (2) 3–4 µm, inamyloid, ornamented with broad low ridges.

Observations   The specific epithet memorializes Gustave Kunze (1793–1851), a famous German botanist who worked in Leipzig.

25   *Ramariopsis kunzei*   About natural size

Clavulina cinerea (Bull. : Fr.) Schroeter   **26**
& *C. cristata* (Holmskjold : Fr.) Schroeter

Identification marks   By maturity the base is usually white to whitish and the branches are gray to brownish gray. The fruiting bodies vary from sparingly branched to much branched and may form dense masses. The tips of the branches are flattened and toothed (cristate) to blunt or irregularly lobed. Dark gray to black "pimples" on some old specimens are produced by a parasite, *Rosellinia clavariae*.

Edibility   Edible and good but often full of sand and difficult to clean.

When and where   Scattered to gregarious on the ground in deciduous, mixed, or coniferous woods under pine; late summer and fall; often abundant in the southern Appalachians. Both species are widely distributed in eastern North America.

Microscopic features   Spores 7.5−9 (10) × 6−8 (9) μm, smooth; variously reported as white or yellow in deposits. Basidia 2-spored.

Observations   This group is a complex of forms and varieties. *C. cristata* has flattened, toothed tips, *C. cinerea* has blunter ones, but intermediates are numerous. *Cinereus* means ash gray; *cristatus* means crested (like a cock's comb).

26   *Clavulina cinerea*   Natural size

Ramaria conjunctipes (Cok.) Cor.   **27**

Identification marks   The tips of the branches are bright yellow, the sides pale salmon orange to yellow orange, and the base nearly white. In age the colors fade to cinnamon buff. The base resembles a cluster of cooked spa-

27 *Ramaria conjunctipes*   Slightly less than natural size

ghetti or small ropes. White rhizomorphs are usually present at the base. Both the odor and taste are mild.

Edibility   We have no information on it but suspect it is nonpoisonous.

When and where   In the South it fruits from August into October in deciduous or mixed woods; it was described from Linville Falls, North Carolina. It also has been reported from New England, the Pacific Northwest, and northern California where it fruits under conifers in the fall.

Microscopic features   Spores 7.5–9 (10) × 4.5–5.5 μm, minutely roughened; yellowish in deposit.

Observations   The ropy base of the fruiting bodies inspired the specific epithet, which means connected foot or stalk. The combination of colors on the tips and sides of the branches along with the distinctive base are diagnostic.

---

28   *Ramaria murrillii* (Cok.) Cor.

Identification marks   The branches are slender and cinnamon brown up to the paler, rounded, buff tips. The base is narrow, sometimes rooting, and typically covered with white hyphae and rhizomorphs. Pinkish to light purplish stains develop on the white areas and on exposed flesh. The flesh is bitter.

Edibility   Not reported.

When and where   Scattered on the ground in woods; summer and early fall; known from the southern Appalachians to the Gulf of Mexico. Murrill found it at Unaka Springs, Tennessee.

Microscopic features   Spores 6.5–9 × (3.5) 4–5.5 μm, light rusty ochraceous in KOH, ornamented with tapering warts; dull ochraceous tan in deposit.

Observations   This is one of a group of species with spiny spores that occurs in the southern United States. The fruiting bodies are drab and unexciting in appearance but the spores are beautifully ornamented. The spores in our collections are a bit wider than usually reported for the species but other aspects of the collections fit the data on *R. murrillii*. W. A. Murrill is commemorated in the specific epithet.

28   *Ramaria murrillii*   Natural size

Ramaria fennica (Karst.) Ricken   **29**

Identification marks   Lilac to violet tones predominate on the upper part of the base and young branches but they fade and are replaced by dull grayish tan to drab tones in old specimens. The fruiting bodies are stout and fleshy. Both the odor and taste are slightly acrid to mild. If a drop of potassium hydroxide (20 percent KOH) is applied to the lilac areas, a bright red color develops which reverts to dull ocher when a drop of acid of similar strength is added.

Edibility   Variously reported to be unpalatable because of a pungent to earthy taste, or mild and presumably edible; not recommended.

When and where   Scattered on the ground in deciduous and coniferous woods; summer and fall; often abundant in the South in upland deciduous woods. It is widely distributed in the Northern Hemisphere.

Microscopic features   Spores 9–12 × 3.5–5.5 μm, minutely roughened; ochraceous in deposit.

Observations   Several varieties have been described, all with the same general coloration but differing in other respects. This variation may account for the differences in reports on its edibility. The specific epithet refers to Finland; this is one of many species described in Europe that also occurs in North America.

29 *Ramaria fennica*   One-half natural size

30 *Ramaria subbotrytis*   About two-thirds natural size

30   *Ramaria subbotrytis* (Cok.) Cor.
   (Rose Coral)

Identification marks This brightly colored coral fungus has salmon orange to salmon pink branch tips and sides when young; the sides become

pinkish ochraceous as the spores mature. The upper part of the base is at first tinged with the color of the branches; the interior is marbled with watery areas. The odor and taste are mild or reminiscent of sauerkraut.

Edibility   Edible, but *R. formosa*, a somewhat similar species, is reported to be poisonous, check specimens carefully.

When and where   Scattered to gregarious or in arcs in deciduous or mixed woods; summer and fall. Chapel Hill, North Carolina is the type locality; the species is widely distributed in eastern North America.

Microscopic features   Spores (8) 9−11 × 3−4 μm, minutely roughened, ochraceous in deposit. Clamp connections absent.

Observations   *R. formosa* differs in having yellow branch tips, clamp connections, and in being fragile and chalky when dried. It also occurs in the South. *Formosus* means handsome or beautiful; *subbotrytis* means resembling *R. botrytis*.

Cantharelloid Aphyllophorales

Fruiting bodies in this group vary from trumpetlike and hollow to resembling a gilled mushroom. The hymenophore is smooth to wrinkled, or if gill-like then the "gills" are usually blunt when young, thick, and often interconnected by low veins at maturity. The technical difference between the two genera included here is that species of *Craterellus* lack clamp connections on the hyphae of the fruiting body while species of *Cantharellus* have them. Of more use to the field mycologist is the fact that species of *Craterellus* generally have funnellike to trumpetlike fruiting bodies that are thin fleshed and hollow and have a smooth to wrinkled hymenophore. In contrast, species of *Cantnarellus* (the genus whose members are commonly called chanterelles) generally have fruiting bodies with a distinct stalk and cap with a smooth, wrinkled, or gill-like hymenophore. Both genera contain a number of excellent edible species.

Craterellus means a large bowl; *Cantharellus* refers to a vase or cup.

Key to Species

1. Fruiting body trumpet- or funnel-shaped; flesh thin; hymenophore smooth to wrinkled 2
1. Fruiting body with a distinct cap and stalk; flesh thin to thick; hymenophore smooth, wrinkled, or gill-like 3
   2. Fruiting body bright to pale orange . . . . 31. *Craterellus odoratus*
   2. Fruiting body gray, grayish brown, or black . . . 32. *Craterellus fallax*
3. Entire fruiting body bright orange-pink to colored like a pink flamingo . . . . 33. *Cantharellus cinnabarinus*
3. Colors of fruiting body yellow, orange, or gray 4
   4. Stalk hollow, cap perforated (opening into the stalk), flesh thin, hymenophore grayish . . . . 34. *Cantharellus tubaeformis*
   4. Not with the above combination of characters 5
5. Hymenophore smooth to only slightly wrinkled . . . .
. . . . 35. *Cantharellus confluens*
5. Hymenophore distinctly gill-like at maturity . . . 36. *Cantharellus cibarius*

31 *Craterellus odoratus* (Schw. : Fr.) Fr.

Identification marks Clusters of bright orange, thin-fleshed "trumpets" arising from a common base are characteristic of this species. In age the trumpets may be complexly folded and intergrown. The clusters may be large and up to 10 cm tall. The inner surface of the trumpets is bright orange; but the smooth exterior, which is where the spores are produced, is paler.

Edibility Edible but almost too pretty to eat; old specimens have an unpleasant flavor.

When and where On the ground in woods; known only from the southeastern United States, uncommon but appearing regularly in certain places. It fruits from April in Florida to September further north.

Microscopic features Spores 7.5–10.5 (12) × 4.5–6 µm; color in deposits not reported. Basidia 4–6-spored.

Observations Schweinitz named the species for its pleasant odor (like violets in his opinion). Since then few reports about it have been published. Not surprisingly, there has been much debate as to whether this species is a *Cantharellus* or a *Craterellus*.

31 *Craterellus odoratus*   About two-thirds natural size

32 *Craterellus fallax* Sm.
(Horn of Plenty)

Identification marks Seen from above, these fruiting bodies often resemble black holes in the ground; on closer inspection they are found to be thin-fleshed, funnel-shaped mushrooms. The lining of the funnel is dark

32 *Craterellus fallax*   Three-fourths natural size

brown to brownish black or grayish brown and is often slightly scaly. The
spores are borne on the smooth to slightly wrinkled exterior of the funnel.

Edibility Edible and popular. The tissue is black when cooked and not exactly attractive.

When and where Scattered or in arcs on the ground in deciduous and
mixed woods; summer and fall; widely distributed and often abundant in
eastern North America.

Microscopic features Spores 12–16.5 × 7–10 μm, salmon buff in deposits. Basidia 2-spored.

Observations Another common name for this species is angel of death.
Fallax means deceptive, this species was long confused with *C. cornu-*
copioides which has white to whitish spores and 4-spored basidia. Several
gray to dark brown species of *Craterellus* occur in the South, all are presumed to be edible. The fruiting bodies of *Urnula craterium* resemble those
of a *Craterellus*, but the former fruits in the spring and is a cup fungus in
which the hymenium lines the cup.

Cantharellus cinnabarinus Schw.   33
(Cinnabar Chanterelle)

Identification marks Flamingo pink to vermillion tones are characteristic
of all parts of young fruiting bodies. In age the caps often fade and become
slightly scaly. The gills are blunt, decurrent, and often anastomosing. The
acrid to peppery taste of the raw flesh is usually absent in cooked specimens.

Edibility Edible and good. The bright color persists in cooked specimens.

When and where Summer and early fall as long as warm, wet weather
persists in deciduous and mixed woods; widely distributed in eastern North
America and common in the South.

33 *Cantharellus cinnabarinus*   About natural size

Microscopic features Spores 7.5–9 × 4–5 µm; pinkish cream in deposit. Basidia 4- and 6-spored.

Observations The bright color fades in dried specimens in a few months. The specific epithet refers to the color of the juice of the dragon's blood tree which is similar in color to cinnabar ore.

---

34 *Cantharellus tubaeformis* (Bull. : Fr.) Fr.

Identification marks The cap is thin fleshed and umber, deep yellowish brown, or grayish brown. The stalk is bright orange near the base and gray near the gills; it is hollow and the cavity extends through the cap, which is said to be perforated. The hymenophore is grayish brown and formed of low, blunt, often interconnected "gills" (more like veins than thin plates). The spores are white in deposit.

Edibility In parts of Europe and Fenno-Scandia it is eaten after parboiling; also reported to be good by some American authors.

When and where Scattered to clustered on damp mossy ground or on very rotten moss-covered logs in mixed forests and in damp coniferous forests; summer and fall; widely distributed in eastern North America.

Microscopic features Spores 9–12 × 7.5–9 µm, white in deposit.

Observations Several species of *Cantharellus* in which the cap is perforated occur in the southern Appalachians. They are separated on spore shape, color of spore deposit, and colors of the fruiting body. The specific epithet means trumpet-shaped, a reference to the form of the mature fruiting bodies. One often gets wet feet when collecting this species since it is usually found in cool wet habitats.

---

35 *Cantharellus confluens* (Berk. & Curt.) Petersen

Identification marks The yellow to orange fleshy caps, smooth hymenophore, and habit of having two or more caps per fruiting body form a distinctive combination. The caps are smooth and neither funnel-shaped nor perforated. Pinkish, apricot, and salmon orange tints are absent at all stages of development.

Edibility   Presumably edible.

When and where   Scattered to gregarious in woods, usually under hardwoods; summer; described from Mexico and to be expected around the Gulf of Mexico and in the southern Appalachians.

Microscopic features   Spores 6–7.5 (10) × 5–6 µm.

Observations   *C. lateritius* also has a smooth to slightly wrinkled hymenophore but it differs in having only one cap per fruiting body and in being light salmon orange to apricot pink. Furthermore, it seems to be more northern in distribution—it occurs as far north as New England. *Confluens* means running together, i.e., cespitose in this case; *lateritius* means brick colored.

34   *Cantharellus tubaeformis*   About natural size

35   *Cantharellus confluens*   Three-fourths natural size

63

36 *Cantharellus cibarius* Fr. : Fr.
(Chanterelle)

Identification marks At all stages the cap and stalk are some shade of pale yellow to egg yolk yellow although the flesh is usually white. Ochraceous stains develop on bruised areas. The gills are distinct, moderately well formed, thick, blunt, and often connected by low veins at maturity. The fruiting bodies are not joined into clusters as a rule, and there is only one cap per stalk. The raw flesh is somewhat peppery but it is mild when cooked.

Edibility Edible and choice; one of the most popular edible mushrooms in the Northern Hemisphere.

When and where Solitary to gregarious on the ground in woods of most types, widely distributed in the forested regions of North America and abundant during warm wet weather in the South.

Microscopic features Spores 7.5–10.5 × 5–6.5 µm; pale buff to ochraceous in deposit.

Observations The color of the spore deposit and that of the fruiting bodies as well as their stature varies considerably in this species. Do not confuse the jack-o'-lantern mushroom *Omphalotus illudens*, which has broad, sharp-edged gills and clustered fruiting bodies associated with wood, with the chanterelles. *Cibarius* means relating to food.

36 *Cantharellus cibarius*   Three-fourths natural size

Agaricales

The majority of fungi of interest to mycophagists belong to this order. The spore-bearing portion of their fruiting bodies is called the hymenophore and is either in the form of tubes or gills. In mycological slang, those species with a tubular hymenophore are called boletes, those will gills, agarics. Fruiting bodies of these fungi are generally short-lived, lasting from only a few hours to a couple of weeks.

Among the major characters used to subdivide this order is the color of the spores in deposits. To evaluate this character, obtain a spore deposit (see p. 6) on white paper, let excess moisture escape for five to ten minutes after the cap is removed, and then judge the color. It may take practice to distinguish the various categories, and making a reference collection of deposits may be useful. If you need to determine whether the spores are inamyloid, amyloid, or dextrinoid, scrape some spores from the deposit onto a white piece of glass or china plate and add a drop of iodine solution and watch for color changes. If they change to black against the light background then they are amyloid; if they become a much deeper rusty orange than the iodine solution, they are dextrinoid; if there is no significant change, they are inamyloid.

Another character that may require some explanation is that of whether the cap and stalk are easily and cleanly separable. In most species with free gills this situation prevails. To judge this character, make a ring with the thumb and index finger of one hand, drop the mushroom to be tested stalk first through the ring, catching the cap so it rests with the gills down on the ring. Then pull gently but steadily on the stalk with your other hand. If the cap and stalk separate cleanly, they will pull apart leaving a rounded "ball" on the top of the stalk and concave "socket" in the cap.

In the following key, for the convenience of the user, we have indicated where certain distinctive genera key out. Some families are easy to distinguish using field characters, others, e.g., the Cortinariaceae and Bolbitiaceae, may be difficult to separate without checking microscopic features.

Key to Families

1. Underside of cap bearing tubes (hymenophore tubular) . . . .
. . . . (p. 67) Boletaceae
1. Underside of cap bearing gills (hymenophore lamellate) 2
    2. Fruiting bodies producing a white to colored latex when gills of young specimens or apex of stalk cut with a knife, typically rather stout (*Lactarius*) . . . . (p. 114) Russulaceae
    2. Not as above 3
3. Cap and gills becoming a black inky liquid at maturity (undergoing autodigestion) (*Coprinus*) . . . . (p. 226) Coprinaceae
3. Not as above 4
    4. Gills free of the stalk, dull pink to dusky red at maturity; spores reddish in deposits; cap and stalk readily separable at maturity
. . . . (p. 198) Pluteaceae
    4. Not as above 5

5. Spores in deposits white to yellow or only lightly colored even in thick deposits; gills often white to pale yellow at maturity 6
5. Spores in deposits distinctly and/or deeply colored: dull red, brown, orange, gray, or black; gills often brown to black at maturity 10
    6. Gills free of the stalk in young specimens 7
    6. Gills attached to the stalk in young specimens 8
7. Cap dry; stalk dry and lacking a volva at the base; spores inamyloid or dextrinoid .... (p. 162) Lepiotaceae
7. Cap tacky to viscid either from the universal veil or beneath remnants of it; remains of an outer veil present at the base of the stalk, on the cap, or in both places; stalk dry or slimy; spores inamyloid or amyloid .... (p. 141) Amanitaceae
    8. Gills waxy in appearance and texture, appearing very clean, often subdistant to distant (see *Laccaria* also) ....
    .... (p. 138) Hygrophoraceae
    8. Not as above 9
9. Fruiting bodies stocky; stalk white or nearly so (in included species) and neither pliant nor fibrous, crumbling when crushed; veils absent .... (p. 114) Russulaceae
9. Not as above .... (p. 172) Tricholomataceae
    10. Gills free of the stalk; stalk and cap cleanly separable; gills white or pink when young becoming chocolate brown as the spores mature .... (p. 233) Agaricaceae
    10. Not as above 11
11. Gills decurrent, distant to subdistant; veils absent; spores olive brown in deposit (*Phylloporus*) .... (p. 67) Boletaceae
11. Not as above 12
    12. Mature gills tinged with dingy red to dusky rose; spores in deposit with distinct red tones .... (p. 201) Rhodophyllaceae
    12. Not as above 13
13. Spores in deposits gray brown, purplish gray, or black 14
13. Spores in deposits rusty orange or some shade of brown but not gray brown or purplish brown 15
    14. Fruiting bodies, especially the stalk, fragile; gills sometimes mottled with shades of gray at maturity ...... (p. 226) Coprinaceae
    14. Fruiting bodies, especially the stalk, pliant to fleshy; gills not mottled at maturity .... (p. 219) Strophariaceae
15. Fruiting bodies typically on, arising from, or adjacent to wood or woody debris 16
15. Fruiting bodies not consistently associated with wood 18
    16. Either cap dry and spores bright rusty orange to rusty ochraceous in deposits (*Gymnopilus*) or spores yellow brown to rusty brown in deposits and cap moist to tacky (*Galerina*) ...
    .... (p. 206) Cortinariaceae
    16. Not as above 17
17. Cap dry to moist to tacky; spores umber brown in deposit; veil ample, membranous (*Agrocybe*) .... (p. 203) Bolbitiaceae
17. Cap tacky to distinctly viscid; spores yellow brown to cinnamon brown in deposits; veil not membranous (*Pholiota*) ....
.... (p. 219) Strophariaceae
    18. Veil, if present, either leaving a braceletlike (tight) annulus or a few wispy fibrils on the stalk; if veils absent then fruiting bodies not particularly fragile .... (p. 206) Cortinariaceae

18. Veil, if present, membranous, often leaving a skirtlike annulus on the stalk; if veils absent then fruiting bodies fragile or cap flesh soon becoming soft .... (p. 203) Bolbitiaceae

Boletaceae

Those members of this family that have a tubular hymenophore are commonly called boletes; they constitute the bulk of the species in the family. A few gilled mushrooms, e.g., *Phylloporus rhodoxanthus*, have such a strong resemblance to certain boletes macroscopically, microscopically, and chemically that they are sometimes placed in the Boletaceae. In the boletes, as in the polypores, the hymenium lines the tubes. Unlike the situation in the polypores, the layer of tubes in the boletes can be separated from the flesh of the cap rather easily at maturity in most species. Bolete fruiting bodies are fleshy and decay within a few days to a week.

Many boletes are excellent for table use. Young firm fruiting bodies are preferable to large old ones that are often soft (and wormy). Remove the tube layer if it is thick, and peel or wipe the cap dry with paper towels if it is slimy, to avoid having slimy to gelatinous cooked boletes. Some species are poisonous and can cause gastrointestinal upsets. The edibility of many species is unknown. Boletes will furnish many good meals if you try only species of proven edibility.

The boletes are a large and relatively poorly known group in the South—one that deserves further study. We have used a conservative classification that places minimal emphasis on microscopic features for the convenience of mushroom hunters. When using the following key, begin with choice *1* if a microscope is available. If not, begin with choice *4*. Some species of *Austroboletus* and *Boletellus* are included in the keys to *Tylopilus* or *Boletus* so that they can be identified if no microscope is available.

Key to Genera and Species

1. Spores ornamented with ridges, striations, pits, or a reticulum 2
1. Spores smooth 4
  2. Cap with dark brown to gray or black fibrillose scales or warts at maturity, flesh turning red then black when exposed to air .... (p. 74) *Strobilomyces*
  2. Not as above 3
3. Spores with longitudinal ridges and/or striations...... (p. 69) *Boletellus*
3. Spores pitted .... (p. 72) *Austroboletus*
  4. Hymenophore in the form of gills ... 86. *Phylloporus rhodoxanthus*
  4. Hymenophore in the form of tubes 5
5. Cap with dark brown, gray, or black fibrillose coarse scales or warts at maturity; flesh becoming black eventually after exposure to air .... (p. 74) *Strobilomyces*
5. Not as above 6
  6. A powdery, bright yellow veil present on young specimens .. .. 37. *Pulveroboletus ravenelii*
  6. Not as above 7

7. Spore deposit bright yellow; tube mouths whitish at first but yellow at maturity; stalk hollow in the base and crumbling when crushed at maturity .... (p. 77) *Gyroporus*

7. Not as above 8

    8. Stalk dry and roughened to squamulose, ornamentation darkening in age to brown, gray, or black .... (p. 78) *Leccinum*

    8. Not as above 9

9. Spore deposit with a purplish, vinaceous, cinnamon, or chocolate brown component; tube mouths never yellow in included species, staining blue only in species with gray to grayish brown hymenophores .... (p. 83) *Tylopilus*

9. Not as above 10

    10. Spore deposit dull cinnamon to medium yellow brown; tube mouths in included species dull yellow, ochraceous, or orange, not staining blue when bruised; stalk with glandular dots or smears (but not slimy overall); if cap is softly fibrillose to scaly, then tube mouths elongated along the radii of the cap .... (p. 90) *Suillus*

    10. Spore deposit olive brown, dull yellow brown, or amber; tube mouths variously colored, staining blue in some species; stalk lacking glandular dots and smears; if cap fibrillose to scaly then tube mouths not strongly elongated along the radii of the cap .... (p. 96) *Boletus*

---

37   *Pulveroboletus ravenelii* (Berk. & Curt.) Murr.

Identification marks A brilliant yellow, cottony to cobweblike veil covers young specimens. In age the remains of the veil may be matted and almost invisible on the cap. At first the cap is bright sulphur yellow and powdery

37 *Pulveroboletus ravenelii*   About one-half natural size

due to the veil; in age it darkens to orange red or brownish red and is tacky to the touch. The flesh and tube mouths slowly stain light blue then brownish where injured. The taste and odor are mild.

Edibility   Reported to be edible.

When and where   Under pines and in mixed woods; summer and fall. It is widely distributed in eastern North America but it is most abundant in the South. It has also been reported from California, Japan, China, Malaya, Singapore, and Borneo.

Microscopic features   Spores 9–10.5 (12) × 5–6 µm. Pleurocystidia 30–46 × 5–9 µm. Cuticle of cap an ixocutis at maturity.

Observations   This species is named for W. H. Ravenel who collected the type specimens along the Santee Canal in South Carolina.

Boletellus

The presence of long ridges and/or striations on the spores is the unifying character of this genus. The spores are some shade of brown in the included species and no veils are present, otherwise the members of this genus have few superficial similarities. *Boletellus* is the diminutive of *Boletus*.

Key to Species

1. Cap shaggy fibrillose to fibrillose-scaly; stalk smooth . . . .
. . . . 38. *Boletellus ananas*
1. Not with both the above features 2
   2. Stalk shaggy reticulate . . . . 39. *B. russellii*
   2. Stalk smooth . . . . 40. *B. chrysenteroides*

Boletellus ananas (Curt.) Murr.   38

Identification marks   The white to pale tan smooth stalk, yellow tube mouths, and purple to dull red fibrillose to scaly cap form a distinctive combination. A zone of pink or purple is often present near the apex of the stalk. Both the tube mouths and the flesh stain blue when injured. A whitish partial veil is present on young specimens but it leaves flaps of tissue along the margin of the cap rather than an annulus.

Edibility   Not recommended.

When and where   Solitary to gregarious at or immediately adjacent to the base of living pines and on the ground in pine-oak woods; late spring to late fall; common and widely distributed in the Southeast at low elevations. The species has also been reported from southeast Asia, Australia, and New Zealand.

Microscopic features   Spores 16–18 (23) × 7.5–9 µm, longitudinally ridged, the ridges transversely striate. Pleurocystidia 37–60 × 9–15 µm. Cuticle of cap of loosely interwoven hyphae arranged in fascicles.

Observations   Curtis described this characteristically southern species from specimens collected by Ravenel. *Ananas* is the name of the genus to which pineapples belong, the relevance of the name to this bolete is unclear.

38 *Boletellus ananas*   About one-third natural size

---

39 *Boletellus russellii* (Frost) Gilb.

Identification marks The dry, squamulose to finely areolate cap; yellow tube mouths; and shaggy, reticulate dull red to red brown stalk form a distinctive combination. The stalk is long in relation to the diameter of the cap and is sometimes slimy near the base.

Edibility Reported to be edible.

39 *Boletellus russellii*   One-half natural size

When and where   Solitary to scattered in association with oaks; summer and fall; widely distributed east of the Great Plains but seldom abundant.

Microscopic features   Spores 16.5–18 (19.5) × 7.5–10.5 μm, ornamented with longitudinal ridges. Pleurocystidia 45–75 × 9–12 μm. Cuticle of cap a trichodermium which collapses in age and forms fascicles of hyphae.

Observations   This distinctive species was named for J. L. Russell (1805–73) a clergyman in Massachusetts whose avocation was botany.

40   *Boletellus chrysenteroides*   Three-fourths natural size

Boletellus chrysenteroides (Snell) Snell   **40**

Identification marks   In appearance this is an "ordinary" bolete in which the cuticle cracks and forms small brown areolae in age. The tube mouths stain blue when bruised. This species cannot be readily identified on the basis of field characters alone and is included to illustrate some of the difficulties one encounters when working with boletes without recourse to a microscope.

Edibility   Reported to be nonpoisonous, but that is not much of a recommendation.

When and where   Summer and fall under hardwoods, often near stumps; widely distributed in eastern North America and sometimes common during warm wet weather. The type collection was gathered near Ithaca, New York.

Microscopic features   Spores 12–15 × 5–6.5 μm, longitudinally grooved. Pleurocystidia 45–66 × 6–10 μm, contents often yellow in KOH. Cuticle of cap a trichodermium.

Observations   Superficially specimens of this species and those of *Boletus chrysenteron* are similar as implied by the specific epithet. However, the spores of *Boletus chrysenteron* are smooth.

Austroboletus

The species in this genus have pitted spores, otherwise, aside from the absence of veils, they have few similarities. The spores are cinnamon brown to olive brown in deposits. The genus name means southern bolete and, except for *A. gracilis*, most species occur in warm climates such as the tropics, subtropics, and adjacent areas.

Key to Species

1. Stalk smooth; cap dark pinkish cinnamon to cinnamon brown; usually growing with white pine and hemlock . . . . 41. *Austroboletus gracilis*
1. Not as above 2
    2. Cap viscid, red when young, fading to yellow in age . . 42. *A. betula*
    2. Cap dry, white to very pale yellow or tinted pinkish orange . . .
    . . . . 43. *A. subflavidus*

---

41 *Austroboletus gracilis* (Pk.) Wolfe

Identification marks Note the narrow, often curved stalk; maroon to cinnamon, velvety cap; unchanging, pinkish to dull grayish pink tube mouths; and mild taste. Yellow tints may develop on the stalk and cap in dried specimens so that little of the cinnamon tones persist. The stalk may be pruinose or reticulate.

Edibility Edible.

41 *Austroboletus gracilis*   About two-thirds natural size

When and where Scattered on humus and rotting wood, often under conifers, hemlock in particular, but also in mixed woods and under hardwoods; summer and fall. It was described from New York; it is widely distributed in northeastern North America and enters the South in the mountains. It has been reported from Japan and Papua New Guinea also.

Microscopic features Spores 13–15 × 6–7 µm, most spores minutely pitted. Pleurocystidia 45–60 × 4–8 µm. Cuticle of cap a trichodermium that collapses in age.

Observations For many years this species has been called *Tylopilus gracilis*; because its spores are ornamented it was recently transferred to *Austroboletus*. We have seen large fruitings of it in North Carolina in woods where white pine, hemlock, rhododendron, and assorted deciduous trees occur together. The specific epithet means slender.

Austroboletus betula (Schw.) Horak   **42**

Identification marks The long, narrow stalk that is dull rosy pink with a yellow, raised, irregular reticulum is noteworthy. Young caps are bright, apple red; they fade to yellow or yellow orange in age but are slimy at all ages. The tube mouths are yellow at first and greenish yellow at maturity. There are no pronounced color changes on bruised flesh or tube mouths.

Edibility Edible but inclined to be soft when cooked.

When and where Solitary to scattered under hardwoods or in mixed forests; summer and fall; widely distributed in northeastern North America south to northern Georgia and west to the Great Lakes. Schweinitz collected it in North Carolina.

Microscopic features Spores 16.5–19.5 × 7.5–10 µm, wall thick and randomly pitted with small pits. Pleurocystidia 40–60 × 7–12 µm. Cuticle of cap an ixotrichodermium which collapses by maturity.

Observations The shaggy stalk supposedly reminded Schweinitz of the bark of a birch tree (genus *Betula*). Until recently this species was often placed in *Boletellus* or *Boletus*.

42 *Austroboletus betula*   One-half natural size

43 *Austroboletus subflavidus*   About one-half natural size

43   *Austroboletus subflavidus* (Murr.) Wolfe

Identification marks   The complex reticulum on the stalk distinguishes this species from the other white to whitish boletes in the region. The caps are whitish to pale yellow or buff and become cracked (areolate) in age. No color changes occur on injured flesh; the flesh is very bitter. Tube mouths may be beaded with clear droplets in young specimens.

Edibility   Undetermined and unappetizing because of the bitter taste.

When and where   Solitary to scattered under oaks and/or pines; summer and fall; Murrill collected it in Gainesville, Florida; it is to be expected in Florida and along the Gulf of Mexico generally.

Microscopic features   Spores 15−17 (20) × 6−7 (8.8) μm, pocked with minute pits particularly over the central portion of the spore, at times appearing reticulate. Pleurocystidia 40−50 × 6−7.5 μm. Cuticle of cap a trichodermium in young specimens.

Observations   The tube mouths are pinkish at maturity and reminiscent of those of some species of *Tylopilus*. *Subflavidus* means somewhat yellowish.

Strobilomyces

By maturity the cap is some shade of gray, brown, or black and the surface is formed into warts or areolate. Cut surfaces of the fruiting body typically turn dull red then black. The tube mouths are pale gray when young and black or nearly so by maturity. The margin of the cap is fused with the stalk in most buttons and leaves a more or less distinct sign of its attachment on the stalk. The spores are dark brown to black in deposits and are, in the North American species, ornamented.

We estimate there are about four species in this genus in eastern North America, none of which are known to be poisonous (but we have little information on two species). The species are easily distinguished by differences in their spore ornamentation, less easily by field characters.

Key to Species

1. Warts on cap small, pointed, and black over the disc by maturity . . . . 44. *Strobilomyces confusus*
1. Warts on cap large, often flattened and dark brown over the disc at maturity . . . . 45. *S. dryophilus*

Strobilomyces confusus Sing.   44

Identification marks   The center of the cap is covered with grayish black to black, firm, small, pointed, erect scales. Above the inconspicuous annulus, the stalk is often reticulate; below it the stalk is fibrillose to woolly. Young tube mouths are whitish; they darken to gray or black in age and stain dull red and finally black where injured.

Edibility   Probably not poisonous. We have no data on this species; the members of the genus, however, are variously rated as good to mediocre. Few mycophagists have reported which species they actually ate.

44   *Strobilomyces confusus*   Three-fourths natural size

When and where Solitary to gregarious under hardwoods and in mixed woods; summer and fall. Although it was described from Florida and is widely distributed in eastern North America, it is most abundant in the Southeast.

Microscopic features Spores 9–11 (12) × 8–10 (12) µm, ornamented with isolated ridges and warts, not reticulate. Pleurocystidia 37–60 × 15–20 µm. Cuticle of cap of fascicles of brown-walled hyphae.

Observations About four taxa in the genus *Strobilomyces* occur in the South, all may be called old man of the woods; the other taxa generally have reticulate spores and blunt, large warts on the cap. *Confusus* means confusing and is appropriate for this species that was long confused with *S. floccopus* in North America.

45   *Strobilomyces dyrophilus*   About one-half natural size

---

45   *Strobilomyces dryophilus* Cibula & Weber (in press)

Identification marks Young specimens and those that remain in the shade are dull grayish pink to pinkish tan. As they age and are exposed to light, shades of dark brown to blackish brown develop on the cap and stalk. The surface of a young cap is fibrillose to woolly but it soon becomes blocked out into broad, low warts that are often flat in age.

Edibility We have no data on this species.

When and where In our experience this species occurs under oak, particularly southern live oak, and fruits in the summer. Its range remains to be determined.

Microscopic features Spores 9.5–12 × 7.5–9 µm, ornamentation in the form of a complete, perfect reticulum to 0.5 µm tall.

Observations The specific epithet means oak-loving, a reference to its apparent mycorrhizal associate. This species is closely related to *S. floccopus* which differs in having a consistently less perfect reticulum on the spores, darker colors in youth, and a more northern distribution.

Gyroporus

The combination of spores that are pale yellow in deposits, pores that are small and nearly circular, and the stalk that is hollow and crumbles when crushed (in age) characterize the genus. In our species the cap is dry and the young tube mouths are whitish until the spores mature. Only one species, *G. cyanescens*, has fruiting bodies that stain blue to violet where injured. These species usually occur under hardwoods or in mixed woods. As far as is known, all the North American species are edible. *Gyroporus* means round pore.

Key to Species

1. Cap and stalk white or tinted with pale seashell pink . . . . 46. *Gyroporus subalbellus*
1. Cap and stalk yellow brown to chestnut brown . . . . . . 47. *G. castaneus*

Gyroporus subalbellus Murr.   **46**

Identification marks The caps are basically white and usually flushed with pale apricot, pink, or yellow. In young specimens the tube mouths are white to whitish; in old ones they are pale, dull yellow. By maturity the stalks are hollow and they are often deeper in color than the caps. No quick color changes occur following injuries, but the caps and stalks may be dotted with brownish to cinnamon-colored spots. The taste is mild.

Edibility The mild taste and lack of known poisonous species in this genus in North America makes it a good subject for experimentation; however, we have no definitive information on its edibility.

When and where Gregarious under oaks and/or pines; late spring into fall. Esther S. Earle (Mrs. F. S.) collected the type specimens in Ocean Springs, Mississippi; this species is common in Florida and widely distributed in the coastal plain.

Microscopic features Spores 9.5–13.5 × 4.5–6 μm. Pleurocystidia not observed. Cuticle of cap of ascending to erect hyphae, a trichodermium.

Observations The specific epithet means somewhat whitish.

46 *Gyroporus subalbellus*   About one-half natural size

One-half natural size

47 *Gyroporus castaneus* (Bull. : Fr.) Quél.
(Chestnut Bolete)

Identification marks Both the cap and stalk are some shade of yellow brown to reddish brown; young tube mouths are whitish, old ones become yellow. Injured areas are unchanging or brownish. By maturity the stalk is hollow.

Edibility Edible; several specimens are needed to make a good serving because of their small size.

When and where In forests and grassy areas with scattered trees, especially oak; spring through fall. Some of the largest fruitings we have seen were in shaded lawns and cemeteries. This common species is widely distributed in eastern North America, but rare in the West; in addition it is widely distributed in the North Temperate Zone and reported from New Zealand and Borneo.

Microscopic features Spores 7.5–11 × 4.5–6 μm. Cuticle of cap a trichodermium which separates into fascicles of hyphae in age.

Observations *G. purpurinus* is similar in stature but is deep purple to purplish red in color. *Castaneus* means chestnut colored.

Leccinum

The combination of the characteristic roughenings on the stalk and the fact that they darken with age separates *Leccinum* from the other boletes. The stalks look as if they had been rubbed with coarse sandpaper to form small tufts of hyphae and squamules.

The genus intergrades with *Boletus* through such species as *B. longicurvipes*.

Although most species of *Leccinum* are relatively ordinary in appearance, there is a wealth of variety at the microscopic level in this genus. The South is particularly rich in species with yellow tube mouths, whereas species with reddish to orange caps are rare to absent in this area. Except for one northern species, *L. atrostipitatum*, those species that have been tested are edible. The name of the genus is derived from an Italian word for fungus.

Key to Species

1. Cap whitish to grayish buff; stalk slender (typically less than 8 mm thick), ornamentation only slightly discoloring by maturity . . . . 48. *Leccinum albellum*
1. Not as above 2
   2. Tube mouths some shade of yellow in young caps 4
   2. Tube mouths whitish to pale gray at first 3
3. Stalk ornamentation black or soon becoming so; cut context staining reddish; cap not areolate . . . . 49. *Leccinum snellii*
3. Stalk ornamentation dark brown to black; cut context soon staining grayish brown to purplish brown; cap becoming areolate by maturity . . . . 50. *Leccinum griseum*
   4. Cap distinctly viscid . . . . see 76. *Boletus longicurvipes*
   4. Cap dry to moist 5
5. Stalk often slightly enlarged in the middle; cap dark brown . . . . 51. *Leccinum crocipodium*
5. Stalk not distinctly enlarged; cap medium yellow brown . . . . 52. *Leccinum rugosiceps*

Leccinum albellum (Pk.) Sing.   48

Identification marks The dry, white to grayish buff or pinkish gray cap; whitish slender stalk; and off-white to pale gray tube mouths remind one of a ghost. Exposed flesh does not change color and the tube mouths are never yellow. Both the odor and taste are mild.

Edibility Probably edible, at least no species of *Leccinum* in this group is known to be poisonous.

When and where Scattered to gregarious under hardwoods, especially oak; summer through fall; widely distributed east of the Great Plains and most abundant in the South.

Microscopic features Spores 13.5–17 (19.5) × 4.5–6 µm. Pleurocystidia often collapsing by maturity, 37–60 × 5–10 µm. Cuticle of cap at first a trichodermium, the cells more or less barrel-shaped, becoming subglobose and separating from each other at maturity.

Observations The presence of subglobose cells in the cuticle and pale colors of the fruiting body are important features of this species. It has been confused with pale specimens of *L. scabrum* and *L. holopus* but these lack subglobose cells in the cuticle and tend to be viscid when wet. The specific epithet refers to the whitish color of the fruiting bodies.

48 *Leccinum albellum*   Just under natural size

49   *Leccinum snellii* Sm. et al.

Identification marks   The cap is dark gray to almost black as is the ornamentation on the stalk. In specimens cut in half longitudinally the flesh near
the apex of the stalk turns orange-red, that at the base becomes blue-green
within fifteen to twenty minutes. Furthermore, the cap typically is slightly
fibrillose and the flesh is firm and mild.

Edibility   Closely related species are listed as edible; we have no reason
to suspect this species is not edible.

When and where   Scattered to gregarious under birch, often abundant;
late summer and fall; widely distributed in eastern North America.

Microscopic features   Spores (16) 18–20 (22) × 5.5–7 µm. Pleurocystidia not observed. Cuticle of cap of appressed hyphae whose cells vary
from oblong to barrel-shaped to occasionally subglobose and may have
brown contents.

Observations   This species was named in honor of Dr. W. H. Snell (1889–
1980) of Brown University who specialized in the study of boletes. *L. scabrum* differs in not changing to reddish in the stalk apex and in lacking
inflated cells in the cap cuticle. Most gray to black species of *Leccinum*
have been called *L. scabrum* at one time or another.

50   *Leccinum griseum* (Quél.) Sing.

Identification marks   Young caps are dull yellowish brown, they darken to
dark brown or grayish brown and become cracked in age. Injured flesh and
tube mouths slowly stain grayish brown to pinkish gray or purplish brown.
The tube mouths are basically pale gray, never yellow.

Edibility European authors rate it edible; we have no information on North American material.

When and where Scattered to gregarious under broad-leaved trees, especially oak, in lawns and grassy places; late summer and fall. Its range in North America is poorly known but it is abundant at times in Mississippi and Michigan after rains.

Microscopic features Spores 15–18 × 6–9 µm. Pleurocystidia persistent, 37–52 × 8–12 µm, contents often brown. Cuticle of cap a trichodermium in which the cells are more or less globose and may separate from each other and be worn away.

Observations The wide spores and abundance of globose to subglobose cells in the cuticle of the cap are important confirming characters. No yellow tints are present. The specific epithet refers to the gray color of the fruiting bodies.

49 *Leccinum snellii*   About one-half natural size

50 *Leccinum griseum*   One-half natural size

51 *Leccinum crocipodium*   One-half natural size

51   *Leccinum crocipodium* (Letellier) Watling

Identification marks   The cap is yellow brown and becomes cracked and somewhat pitted in age. Frequently the stalk is enlarged in the middle. In contrast to *L. griseum* the tube mouths of this bolete are bright yellow. Specimens cut in half longitudinally show that the context darkens when exposed to air.

Edibility   Some European authors rate it as edible, others say members of this group are suspect. We have no data on North American material and advise caution or avoidance.

When and where   Scattered to gregarious where oak is present; to be expected in the coastal plain especially in northern Florida and the Mississippi Delta; late summer and fall.

Microscopic features   Spores 12–15 × 5–6.5 µm. Pleurocystidia 36–60 × 9–13 µm, often collapsed by the time the spores are mature. Cuticle of cap a trichodermium, by maturity the hyphae clustered in fascicles, the cells not or only slightly inflated.

Observations   We are using a broad species concept for a complex group much in need of further study. The important characters are the yellow tube mouths, yellow brown areolate cap, and lack of greatly inflated cells in the cuticle of the cap. The specific epithet is generally interpreted to mean saffron footed.

52   *Leccinum rugosiceps* (Pk.) Sing.
(Rough Cap Bolete)

Identification marks   Important features include the yellow tube mouths, dull brownish yellow to crust brown cap, and yellow to pale orange yellow stalk ornamented with scurflike spots. The tube mouths are unchanging to dull reddish when injured as is the flesh. By maturity the surface of the cap is pitted to finely wrinkled and often finely cracked The ornamentation on the stalk darkens with age.

Edibility   Edible and good according to reports.

When and where   Scattered on the ground in thin oak woods and oak-shaded lawns; late spring into fall; common in the South and rare to sporadic in the North although it was described from specimens gathered at Port Jefferson, New York.

Microscopic features   Spores 15–17 (20) × 5–6 µm. Pleurocystidia rare, 36–48 × 7.5–10.5 µm. Cuticle of cap a trichodermium with a mixture of globose, subglobose, and ellipsoid cells.

Observations   The specific epithet refers to the wrinkled or rough cap. Several other species have similarly colored fruiting bodies but differ in anatomical details and spore characters. All are presumed to be edible until proven otherwise.

52   *Leccinum rugosiceps*   About two-thirds natural size

Tylopilus

The spores are purplish brown to cinnamon brown or chocolate brown in deposits. There are neither veils nor slimy to viscid caps in our species. Species with dark grayish brown hymenophores at maturity may stain waxed paper greenish blue and stain blue on the tube mouths but otherwise no changes to blue as seen in *Boletus* occur in the genus. The tube mouths in many species are pinkish tan, brown, or gray at maturity, seldom yellow as is common in *Boletus*.

Those species with mild flesh that have been tested are generally rated edible but many species in this genus are unpalatably bitter. *Tylopilus* means bumpy cap. Two species of *Austroboletus* with pinkish brown to cinnamon spores are included in the key.

Key to Species

1. Cap shaggy-fibrillose, color a mixture of pale yellow, ochraceous, and seashell pink .... 53. *Tylopilus conicus*
1. Cap smooth to areolate 2
    2. Cap and stalk white or nearly so, sometimes pale tan or tinged pale yellow to pale pinkish orange 3
    2 Cap, and often stalk, deeper in color: purple, brown, orange, gray, etc. 4
3. Reticulum on stalk complex, deep; cap often areolate by maturity .... see 43. *Austroboletus subflavidus*
3. Reticulum on stalk low, simple; cap smooth in age .... 54. *Tylopilus rhoadsiae*
    4. Cap orange at first, fading to brown in age .... 55. *Tylopilus balloui*
    4. Not as above 5
5. Cap and stalk dark gray to black; context becoming red then black when injured .... 56. *Tylopilus alboater*
5. Not as above 6
    6. Hymenophore gray or brown at maturity 7
    6. Hymenophore whitish at first becoming pale pinkish tan at maturity 8
7. Fruiting bodies staining greenish blue where in contact with waxed paper and also staining the waxed paper; stalk and cap sooty gray .... 57. *Tylopilus fumosipes*
7. Fruiting bodies not staining greenish blue as above; stalk and cap tawny to yellow brown .... 58. *Tylopilus tabacinus*
    8. Taste of flesh mild 9
    8. Taste of flesh bitter .... 59. *Tylopilus plumbeoviolaceus*
9. Cap becoming granulose to finely areolate at maturity; stalk slender; growing with hemlock .... see 41. *Austroboletus gracilis*
9. Cap remaining smooth at maturity; stalk often medium to stout in relation to cap; growing with hardwoods .... 60. *Tylopilus indecisus*

---

53   *Tylopilus conicus* (Berk. & Curt.) Beards.

Identification marks   Delicate shades of yellow and seashell pink predominate on the shaggy-fibrillose, broadly conic caps which are fringed along their margins at first. In age the cap becomes pitted as the fibrils pull apart and are compacted. The stalk is slender. The pores are pinkish tan and unchanging when bruised.

Edibility   Unknown; however, the flavor of the flesh is mild.

When and where   Scattered in low pine forests which may be flooded part of the year; summer and fall; North Carolina to Florida. It was described from a collection made by Ravenel. Subsequent sightings of it were noted in 1934 and 1947 when it was characterized as an extremely rare and little-known species. We suspect in its characteristic habitat it may be fairly common but that mycologists have been uncommon in that habitat.

Microscopic features   Spores 15–21 (23) × 4.5–6 µm. Pleurocystidia 41–68 × 5–9 µm, often collapsing. Cuticle of cap a loose trichodermium grading into a lattice; some gelatinization present in age.

Observations   A proposal has been made to place this species in *Mucilopilus* as *M. conicus* based on the presence of gelatinous material in the cuticle of the cap.

53   *Tylopilus conicus*   About two-thirds natural size

54   *Tylopilus rhoadsiae*   About one-half natural size

Tylopilus rhoadsiae (Murr.) Murr.   54

Identification marks   The combination of whitish cap, whitish distinctly reticulate stalk, and pale grayish pink tube mouths at maturity is distinctive. The reticulum is low unlike that of *Austroboletus subflavidus*. The context does not change color when it is injured. Of five people who tasted bits of the same raw cap, two pronounced it mild, three said it was very bitter.

Edibility   There will be a difference of opinion on the desirability of this species as an esculent as long as some people taste bitterness in it and others do not; we have no information on its toxicity or lack thereof.

When and where   Under pines, oaks, and in open pine-oak woods; summer; common in northern Florida and to be expected throughout the coastal plain; reported from as far north as New Hampshire.

Microscopic features   Spores 12–13.5 × 3.5–4.5 µm. Pleurocystidia 39–46 × 7.5–10.5 µm. Cuticle of cap a tangled trichodermium of narrow hyphae.

Observations   *T. intermedius* is similar in color and stature but has only a faint reticulum near the apex of the stalk and is apparently northern. Murrill described *T. rhoadsiae* from specimens collected by Louise and Arthur S. Rhoads near Melrose, Florida and named it for Mrs. Rhoads.

---

55   *Tylopilus balloui* (Pk.) Sing.

Identification marks   The bright bittersweet orange cap contrasts with the off-white, smooth to faintly reticulate stalk and whitish tube mouths. Injured tube mouths stain brown. Raw flesh is mild to slightly unpleasant in taste.

Edibility   Edible; reports on its quality vary.

When and where   Scattered to gregarious under hardwoods and/or pines; summer and fall; widely distributed in the South and occurring at least as far north as Massachusetts and Ohio.

Microscopic features   Spores 7.5–9 (11) × 3.5–4.5 µm. Pleurocystidia 30–45 (78) × 9–18 µm, contents often ochraceous in KOH. Cuticle of cap of tangled interwoven hyphae sometimes grouped into fascicles.

Observations   In the North one might think this was a *Leccinum* because of the orange cap, but the ornamentation, when present on the stalk, does not darken in age. Old caps may be dull orange to brownish. W. H. Ballou (1857–1937) did much of his mushroom hunting in and around New York City and northern New Jersey.

55   *Tylopilus balloui*   Three-fourths natural size

56 *Tylopilus alboater*   One-third natural size

Tylopilus alboater (Schw. : Fr.) Murr.   **56**

Identification marks   This fleshy, firm, large bolete can easily be mistaken for a rock. The dry cap and stalk are dark gray to black. No reticulum is present on the stalk. The pores are whitish at first and become dull rosy pink in age; when injured they stain dingy purplish red then black. The context is mild, whitish at first but soon changes to dull flesh pink then black when exposed.

Edibility   Reported to be edible and of good flavor. The color changes of injured flesh might be startling to the unprepared.

When and where   Scattered to gregarious under hardwoods, oak in particular; summer and early fall. Described from North Carolina, it occurs from New England to northern Florida west to Mississippi in North America and has also been reported from Malaya and Singapore.

Microscopic features   Spores 9–12 × 3.5–5 µm. Pleurocystidia (22) 45–67 × 7–15 µm, contents dull brown in KOH. Cuticle of cap at maturity of loosely interwoven hyphae, the end cells often grouped in fascicles.

Observations   The specific epithet means white and black. This species is an example of the principle that good looks and good eating are not necessarily related in the fungi.

Tylopilus fumosipes (Pk.) Sm. & Thrs.   **57**

Identification marks   The dark brown dry cap soon becomes areolate, exposing the whitish context which sometimes becomes reddish in the cracks. Young tube mouths are whitish, in age they become deep brown; injured mouths quickly stain deep blue. Little or no ornamentation is present on the stalk.

Edibility   We have no information on it; not recommended.

When and where   Solitary to scattered in woods and shaded parks and grassy areas; widely distributed east of the Great Plains but seldom abundant; summer and fall; Peck gathered the type collection at Port Jefferson, New York.

Microscopic features   Spores 9–12 (15) × 5–7 μm. Pleurocystidia 50–75 × 10–12 μm, contents dull yellow to rusty brown in KOH. Cuticle of cap a trichodermium that is separated into patches by expansion of the cap.

Observations   This is one of about a half-dozen species with similar field characters. None is reported to be good for table use. Several of them stain waxed paper green to bluish green. As applied to this species the specific epithet refers to the smoky color of the stalk.

57   *Tylopilus fumosipes*   Two-thirds natural size

58   *Tylopilus tabacinus* (Pk.) Sing.

Identification marks   All surfaces of the fruiting body are some shade of bright yellow brown, tawny, or tobacco brown. The cap is dry and smooth on the surface; its flesh is soft, white becoming pale purplish buff when exposed; and has a slightly nauseous unpleasant to faintly bitter taste. The stalk is reticulate. Injured tube mouths are unchanging to slightly darker than uninjured ones.

Edibility   Not recommended because of the lack of reliable data on its edibility.

When and where   In shaded lawns, parks, and thin woods; apparently associated with oaks and/or pines. Peck described the species from specimens collected in Alabama by Underwood. It is widely distributed in the South and has been reported as far north as Rhode Island. It fruits in the summer and early fall.

Microscopic features   Spores 10.5–12 (13.5) × 3.5–4.5 μm. Pleurocystidia 35–45 (75) × 8–11 (15) μm. Cuticle of cap a trichodermium.

Observations   These fruiting bodies can be quite large. *Tabacinus* means tobacco colored or pale brown.

58 *Tylopilus tabacinus*   One-half natural size

Tylopilus plumbeoviolaceus (Snell & Dick) Sing.   **59**

Identification marks A fresh young specimen is most attractive with its deep dusky violet velvety cap, whitish tube mouths, and violet stalk that is smooth or reticulate only near the apex. In age the cap becomes browner; the violet to deep purple tones persist longest on the stalk. At maturity the tube mouths are pinkish tan. The flesh is white and extremely bitter.

Edibility Unpalatable because of the bitter taste.

When and where Gregarious to scattered under oaks; summer and fall; widely distributed east of the Great Plains but most abundant in the South.

Microscopic features Spores 9–10.5 (13) × 3–4 µm. Pleurocystidia 37–52 × 8–12 µm, hyaline or in age contents yellow brown. Cuticle of cap a trichodermium of narrow hyphae whose cells (as viewed in KOH) contain brown pigment globules.

Observations The specific epithet is compounded from words meaning leaden gray and violet. *T. rubrobrunneus* also has bitter flesh; however, its stalk typically lacks a reticulum, the cap is reddish brown, and the fruiting body larger.

59 *Tylopilus plumbeoviolaceus*   About one-third natural size

60 *Tylopilus indecisus*   One-half natural size

60 *Tylopilus indecisus* (Pk.) Murr.

Identification marks These boletes have white mild flesh that is firm in young specimens. The cap is ochraceous brown when young and becomes dull cinnamon in age. The tube mouths are pale pinkish tan and stain pinkish brown to rusty ochraceous when injured. Typically a reticulum is present near the apex of the stalk, but it is often inconspicuous to absent.

Edibility Edible; firm young buttons are best. Some closely related species are very bitter.

When and where Scattered to clustered under hardwoods, especially oaks, often in lawns and parks as well as forests; summer and fall; widely distributed in eastern North America and one of the common summer species in the South. Peck gathered the type collection near Menands, New York.

Microscopic features Spores (10.5) 12–15 (16.5) × 4–5 µm. Pleurocystidia 30–48 × 8–11 µm, contents hyaline to dull orange in KOH. Cuticle of cap a tangled trichodermium of narrow hyphae.

Observations No reticulum was present on most of our collections identified as *T. indecisus* and we are undecided about the significance of this variation. Peck's description states that the stalk is "reticulated at the top." *Indecisus* means undecided.

Suillus

This is an easy genus to recognize once you have a bit of experience with it, but a hard one to characterize on paper. In addition to having spores that are some shade of yellow brown to olive brown in deposits, one or more of the following features must be present for a bolete to qualify for inclusion in *Suillus*: tube mouths elongated along the radii of the cap (not round); if tube mouths round then the stalk bears (but is not covered by) sticky dots and/or smears (glandular dots) that are often brown at maturity; cap either

dry and fibrillose or viscid to some degree; cystidia on the hymenophore typically in clusters.

Most species in this genus whose edibility has been tested are
edible, but occasional unpleasant reactions have been' reported.
Young firm specimens are preferred for eating. Wipe off or peel off
the sticky cuticle (and the tube layer if it is thick) before cooking
them. All the species of *Suillus* probably form mycorrhizal associations with various species of trees, particularly conifers. *Suillus*
means pertaining or belonging to swine and was a name used in
classical times for a kind of fungus.

Key to Species

1. Cap dry and fibrillose, partial veil present on buttons 2
1. Cap viscid, either glabrous or fibrillose; partial veil present or ab-
   sent 3
   2. Cap rosy red to purplish red; associated with white pines . . . .
      . . . . 61. *Suillus pictus*
   2. Cap apricot orange to dull ochraceous yellow; associated with
      southern 2- and 3-needle pines . . . . 62. *S. decipiens*
3. Cap fibrillose-scaly when young, the scales immersed in slime in
   age . . . . 63. *S. hirtellus*
3. Cap glabrous at all stages of development 4
   4. Veil present on buttons, leaving a distinct annulus on the stalk
      of mature specimens . . . . 64. *S. salmonicolor*
   4. No veil present on fruiting bodies 5
5. Cap milky white to pale yellow; numerous glandular dots and
   smears present on stalk . . . . 65. *S. placidus*
5. Cap pale to medium brown; stalk with few, if any, glandular dots
   . . . . 66. *S. brevipes*

Suillus pictus (Pk.) Sm. & Thrs.   61

Identification marks   This attractive bolete has dry, purple red to rosy red
fibrils over the cap and on the stalk below the annulus; yellow tube mouths
elongated along the radii of the cap; a mild taste; and yellow flesh. When
young, a creamy yellow to reddish fibrillose partial veil is present; it leaves a
cottony, often grayish zone on the stalk. Glandular dots are absent.

Edibility   Edible and considered good by many mycophagists including
both squirrels and people.

When and where   It is associated with eastern white pine; in the South it
is common only in the mountains where it fruits from summer into fall and
can often be collected in quantity.

Microscopic features   Spores (8) 9–11 × 3–4.5 µm. Pleurocystidia
40–60 (75) × 6–9 µm, scattered or in small clusters. Cuticle of cap of
loose fibrils.

Observations   The specific epithet means the painted one. When tube
mouths are described as "boletinoid" the type of tube mouth characteristic
of this species is what is meant. We have seen specimens of this species
stashed in trees, apparently placed there to dry by squirrels.

61 *Suillus pictus* Two-thirds natural size

62 *Suillus decipiens* Two-thirds natural size

62 *Suillus decipiens* (Berk. & Curt.) Kuntze

Identification marks The caps are basically apricot orange to pinker or dull yellow and coated with soft dry fibrils and scales. The veil is white at first and often stains gray in age; it leaves a cottony zone on the stalk. No glandular dots are present on the stalk. The tube mouths are large and elongated along the radii of the cap.

Edibility Edible.

When and where Summer into winter in dry pine woods and pine-oak woods; widely distributed and common in the coastal plain and Florida. Curtis collected it in South Carolina.

Microscopic features Spores 7.5–10.5 × 3–4.5 μm. Pleurocystidia 48–52 × 6–8 μm, in fascicles. Cuticle of cap of dry, loosely tangled hyphae and fascicles of hyphae.

Observations Small fruiting bodies that never expand are sometimes found with normal ones. *Decipiens* means deceiving, i.e., closely resembling another species. *Boletinellus merulioides* in this case, but the resemblance is not great to trained eyes.

Suillus hirtellus (Pk.) Kuntze   63
(Hairy Bolete)

Identification marks Young caps are covered with short tufts of yellowish to apricot-colored fibrils. As they age the scales and fibrils darken to the color of cinnamon, mat down, and are embedded in slime. The cap margin is typically grayish and translucent. We have no proof, however, that there is ever a true veil. At first the glandular smears and dots on the stalk are yellowish, by maturity they become brown to brownish red.

Edibility Not reported.

When and where Look for it under 2- and 3-needle pines from summer until the end of the growing season in the coastal plain and Piedmont. It was described from New York.

Microscopic features Spores 8–10 × 3–4 μm. Pleurocystidia in fascicles, 27–48 × 4.5–7.5 μm. Cuticle of cap of radially arranged hyphae and fascicles of hyphae in a gelatinous matrix.

Observations *S. hirtellus* is treated here as a collective species characterized by the viscid cap with embedded scales, conspicuous glandular smears on the stalk, lack of blue stains on bruised areas, and lack of a distinct veil. *Hirtellus* means covered with small hairs; Peck merely translated the scientific name when he proposed the vernacular name.

63 *Suillus hirtellus*   About three-fourths natural size

64 *Suillus salmonicolor* (Frost) Halling

Identification marks Slimy, dull yellow to cinnamon, or, in age, olive brown to dark brown caps; orange yellow to orange flesh; and small, peach-colored to pale salmon tube mouths are important features. A baggy veil is present on young specimens and leaves a bandlike annulus on the glandular-dotted stalk. The odor is slightly fragrant.

Edibility Not reported.

When and where Scattered or in clusters under 2- and 3-needle pines from late summer in the North until the onset of freezing weather in the South. This is a common species in lawns under pines. It was described from Vermont and is common and widely distributed along the Atlantic coast west into Texas.

Microscopic features Spores 7.5–9 (10) × 3–3.5 μm. Cystidia 34–60 × 10–13 μm often in fascicles. Cuticle of cap an ixotrichodermium that collapses into an ixocutis.

Observations This is a species that has been overlooked by professionals and amateurs alike for many years. There are several species that need to be reexamined and compared with *S. salmonicolor*, in particular *S. cothurnatus* that differs, if it is significant, primarily in having yellow rather than salmon-colored tube mouths. Both variants are common along the Gulf Coast and deserve careful study. *Salmonicolor* refers to the salmon-colored portions of the fruiting bodies.

64 *Suillus salmonicolor*   About two-thirds natural size

65 *Suillus placidus* (Bonorden) Sing.

Identification marks Milky white, slimy, glabrous caps are characteristic of specimens in prime condition. In age the caps may darken to pale yellow and the slime darkens to light brown or gray. Copious sticky dingy pink to dull watery-purple dots and smears nearly cover the stalk. There are no veils.

65 *Suillus placidus*   About natural size

Edibility Presumably edible, at least we have no information to the contrary.

When and where Scattered to gregarious under eastern white pine from summer until the end of the growing season. It occurs from the southern Appalachian Mountains north into Canada.

Microscopic features Spores 7−9 × 2.5−3.5 µm. Hymenial cystidia 49−60 × 6−9 µm, typically in fascicles. Cuticle of cap an ixotrichodermium that soon collapses.

Observations We include this species as another reminder that many species with typically northern distribution patterns can be found in the Appalachian mountains. The button in the photograph did not develop properly and has the colors of an old specimen. *Placidus* means smooth or pleasing.

Suillus brevipes (Pk.) Kuntze   66
(Short-stemmed Bolete)

Identification marks These stout, typically short-stalked boletes have smooth, slimy, brown caps. There are no veils. Glandular dots are often absent from the stalk and are never well developed. The flesh is white, firm, and unchanging when exposed.

Edibility Edible and choice. Peel off the slimy elastic cuticle of the cap before cooking.

When and where Scattered to gregarious or cespitose under pines, especially 2- and 3-needle pines; late fall through the winter in the South; fall in the North. It is widely distributed in North America and one of the best edible boletes.

Microscopic features Spores 7.5−9 (12) × 2.5−3.5 µm. Pleurocystidia in fascicles, 35−50 (60) × 4.5−9 µm. Cuticle of cap an ixocutis.

Observations Firm young buttons are excellent when cooked; in age the flesh tends to become soft. The specific epithet means short footed, an appropriate name for this species.

66 *Suillus brevipes*   Just under natural size

Boletus

Boletus is the largest genus in the family and as such is the most difficult to characterize. The spores are smooth and olive brown, amber brown, or grayish brown in deposits. The pores are not boletinoid as in some species of *Suillus*. In most species the tubes, but not necessarily the tube mouths, are white, pale gray, or yellow. Bruised areas may change to blue, red, or rusty brown. There are no veils. If the stalk is squamulose, as in *Leccinum*, the ornamentation does not become brown to black in age.

Many species are edible; some, particularly those in which the tube mouths and tube walls differ in color and stain blue when injured, may cause gastrointestinal upsets. There is no information on the edibility of many species. Some species of *Austroboletus* and *Boletellus* are included in the following key because these genera cannot be readily distinguished from *Boletus* without checking the spore ornamentation. The name *Boletus* was used in classical times for a type of edible fungus; in Greek it means clod.

Key to Species

1. Stalk coarsely reticulate and typically at least twice as long as the cap is wide by maturity 2
1. Not as above 3
    2. Cap dry, dull brownish red to brownish yellow ....
        .... see 39. *Boletellus russellii*
    2. Cap slimy and bright red when young (see 71. *Boletus frostii*)
        .... see 42. *Austroboletus betula*
3. Cap coarsely fibrillose-scaly, rosy pink at first fading to tan or dingy white .... see 38. *Boletellus ananas*
3. Not as above 4
    4. Cap, stalk, and tube mouths yellow; cap and stalk slimy when fresh, not staining blue; stalk with a white cottony base .....
        .... 67. *Boletus curtisii*
    4. Not as above 5

5. Fruiting body apparently parasitic on that of a *Scleroderma* (earthball) .... 68. *Boletus parasiticus*

5. Not as above 6

    6. Cap bright yellow, surface soft and powdery; tube mouths red to reddish brown; stalk yellow; typically fruiting on or at the base of pines .... 69. *Boletus hemichrysus*

    6. Not as above 7

7. Tube mouths red to orange; walls typically yellow; bruised areas often staining blue 8

7. Not as above; if tube walls yellow then mouths also yellow; bruised areas staining blue or not 12

    8. Cap whitish to pale grayish tan .... 70. *Boletus piedmontensis*

    8. Cap some shade of red, rose, or brown 9

9. Cap viscid or dry, bright red; stalk reticulate 10

9. Cap dry, dusky rose or brown; stalk not reticulate 11

    10. Stalk ornamented with a heavy, complex reticulum .... 71. *Boletus frostii*

    10. Stalk smooth or usually with a low, simple reticulum .... 72. *Boletus flammans*

11. Cap dusky rose to red .... 73. *Boletus bicolor*

11. Cap some shade of brown .... 74. *Boletus erythropus* & *Boletus hypocarycinus*

    12. Cap and stalk yellow, often with rusty red streaks on the cap; injured areas quickly turning blue .. 75. *Boletus pseudosulphureus*

    12. Not as above 13

13. Cap viscid when fresh; stalk squamulose .... 76. *Boletus longicurvipes*

13. Not as above 14

    14. Stalk reticulate 15

    14. Stalk smooth or if ornamented not reticulate 18

15. Cap gray; stalk gray or gray with a yellow base .... 77. *Boletus griseus*

15. Not as above 16

    16. Flesh yellow, taste bitter; cap some combination of olive, gray, and yellow .... 78. *Boletus retipes*

    16. Not as above 17

17. Cap yellow brown to dark brown; growing under hardwoods or in mixed woods .... 79. *Boletus variipes*

17. Cap cinnamon brown to chestnut brown; growing under pines or in mixed woods .... 80. *Boletus pinophilus*

    18. Cap brown to purple brown 19

    18. Cap rose to red 22

19. Cap areolate at maturity .... see 40. *Boletellus chrysenteroides*

19. Not as above 20

    20. Tube mouths brilliant yellow to greenish yellow when young 21

    20. Tube mouths whitish to pale tan when young ... 81. *Boletus affinis*

21. Stalk slender, often tinged rose; cap greenish brown .... 82. *Boletus viridiflavus*

21. Stalk stout, never tinged with rose; cap deep brown .... 83. *Boletus caespitosus*

    22. Cap areolate at maturity .... 84. *Boletus fraternus*

    22. Not as above 23

23. Stalk yellow overall, seldom over 1.5 cm thick .. 85. *Boletus campestris*

23. Stalk red like the cap over the lower half to two-thirds, typically 1–3 cm thick .... see 73. *Boletus bicolor*

67　*Boletus curtisii* (Berk.) Sing.

Identification marks　This bright yellow bolete in which both the cap and stalk are slimy might be mistaken for a *Suillus*; however, the stipe lacks glandular dots, the pores are not elongated radially, and there is no veil. The cottony to marshmallowlike white zone at the base of the stalk is characteristic of the species. Neither the flesh nor the pores stain blue when injured.

Edibility　We have no information on it.

When and where　Scattered to gregarious in pine woods or mixed woods; summer and fall; rare to occasional; described from South Carolina and known from southern New England to the Carolinas, west into Texas, and north into Indiana.

Microscopic features　Spores (9) 10.5–12 (15) × 4–5 (6) μm. Pleurocystidia scattered, 45–60 (70) × 7–9 μm. Cap cuticle an ixocutis at maturity, the hyphae indistinct.

Observations　The species is named for M. A. Curtis. It is one of the few boletes with both a viscid stalk and a viscid cap.

67　*Boletus curtisii*　　　　　　　　　　　Three-fourths natural size

68　*Boletus parasiticus* Bull. : Fr.

Identification marks　This is the only bolete regularly associated with earthballs (*Scleroderma citrinum*) in North America. Several fruiting bodies of the bolete seem to arise from a single shrunken and deformed earthball like petals around the center of a daisy. The cap is viscid, the flesh soft and yellow, and the odor and taste mild. In age the tubes become soft and gelatinous.

Edibility Nonpoisonous but the soft texture and small size are not appealing.

When and where Late summer into fall in moist mixed woods; widely distributed in eastern North America but most abundant in the southern Appalachians.

Microscopic features Spores 12–15 × 4.5–5.5 µm. Pleurocystidia (35) 50–60 × 7–11 (13) µm. Cuticle of cap a tangled ixotrichodermium.

Observations The specific epithet refers to the apparent parasitic habit of this species. In an area where it is fruiting, both parasitized and healthy specimens of the *Scleroderma* can often be found.

68 *Boletus parasiticus*   About natural size

Boletus hemichrysus Berk. & Curt.   **69**

Identification marks When Berkeley and Curtis described this species, they specified a bolete with a bright golden yellow floccose cap covered with yellow powder, rich red brown tube mouths, decurrent tubes, a short stalk that is yellow tinged with red, and yellow basal mycelium. The specimens in our illustration, although young, seem to have all the required features. The flesh in the top of the stalk turned blue when exposed to air.

Edibility We have no information on it; not recommended.

When and where Adjacent to, on, or near living trees or stumps of pine, especially southern pines; summer. It occurs at least from Nova Scotia to Florida and west to Arkansas. In the type collection, gathered by Ravenel in South Carolina, the specimens were attached to tree roots.

Microscopic features Spores 6–9 × 2.5–3 µm. Hymenial cystidia inconspicuous, approximately 20–30 × 4.5–6 µm. Cuticle of cap a thick region of loosely interwoven, anastomosing hyphae.

Observations Little is known about this colorful and distinctive bolete; it seems to appear regularly and is often abundant along the Gulf coast. The specific epithet means half yellow.

69 *Boletus hemichrysus*   Natural size

70   *Boletus piedmontensis* Grand & Sm.

Identification marks The smooth, dry, off-white to pinkish tan or light grayish olive cap; orange red to deep red tube mouths that stain blue when bruised; and stalk that is neither bulbous nor strongly reticulate are distinctive. The whitish flesh is thick and slowly stains blue when injured.

Edibility Not recommended; we have no data on this species; however, some related ones are poisonous.

70 *Boletus piedmontensis*   One-third natural size

When and where Scattered in deciduous or mixed woods; summer and fall. Its range is not well known; it was described from the Piedmont region of North Carolina and we have collected it in Georgia and along the Gulf of Mexico in Mississippi.

Microscopic features Spores (9) 10–12 × 3.5–4.5 µm. Pleurocystidia 35–52 × 7–9 µm. Cuticle of cap composed of loosely interwoven hyphae whose tips are sometimes grouped into fascicles.

Observations The specific epithet means relating to the Piedmont.

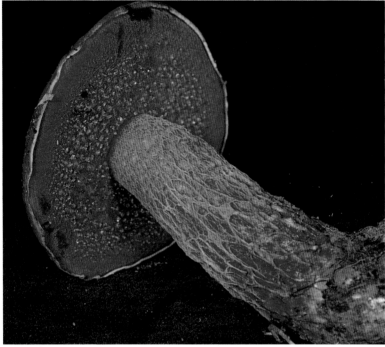

71 *Boletus frostii*   About natural size

Boletus frostii Russell in Frost   71

Identification marks The bright, apple red cap is viscid especially in young specimens. There may be drops of amber liquid on the tube mouths. The tube mouths are red at all stages. A prominent and complex reticulum covers the stalk and is red or occasionally yellow. Blue stains develop on injured areas.

Edibility Although it is said to be edible, we do not recommend eating any bolete with red, orange, or brown tube mouths that stain blue.

When and where Scattered under hardwoods especially oak, and in oak-pine woods; widely distributed east of the Great Plains and common in summer and fall along the Atlantic and Gulf coasts.

Microscopic features Spores 11–15 × 4–5 µm. Pleurocystidia 24–38 × 8–12 µm. Cuticle of cap an ixocutis.

Observations This is one of the most attractive species in the genus. J. L. Russell (1805–73) named this species for C. C. Frost (1805–80); in the same paper Frost dedicated *Boletellus russellii* to Russell.

72   *Boletus flammans* Dick & Snell

Identification marks   The deep red color of the cap, stalk, and tube mouths make this a conspicuous bolete. A faint to distinct reticulum is present on the stalk, particularly near the apex. All parts stain blue when injured. The cap is dry to the touch.

Edibility   Not known and not recommended. *B. rubroflammeus*, a closely related species, was the apparent cause of a severe gastrointestinal upset and the two species are difficult to distinguish in the field.

When and where   Scattered under hemlock, white pine, and/or spruce; late summer and fall. Its known range includes Nova Scotia, New England, North Carolina, and Texas.

Microscopic features   Spores 10.5–12 × 3.5–5.5 µm. Pleurocystidia 50–60 × 5–7 µm, rare. Cuticle of cap a trichodermium that becomes a lattice in age with the end cells in tufts.

Observations   *B. rubroflammeus* seems to be associated with hardwoods and has a cap cuticle composed of tightly interwoven hyphae. *Flammans* refers to the flaming red colors of the fruiting bodies.

72 *Boletus flammans*   One-half natural size

73   *Boletus bicolor* Pk.

Identification marks   Several species share the attributes of *B. bicolor*: red dry caps, yellow tube mouths that stain blue when injured, and little or no reticulum on the stalk. *B. bicolor* is distinguished by the combination of medium-sized fruiting bodies with small round pores, rusty red to purple red stalks, and apple red rather than orange red caps. The caps often fade in age. Two varieties are shown here, var. *bicolor* with yellow tube mouths and var. *borealis* with red to orange red tube mouths.

Edibility   Not recommended; although some people eat and enjoy var. *bicolor*, others cannot tolerate it. Furthermore, it is difficult to make accurate identifications in this group on field characters alone.

73a *Boletus bicolor* var. *bicolor*   One-half natural size

When and where   In mixed forests or under hardwoods, common and widely distributed in eastern North America in the summer and fall. Var. *bicolor* is more northern and montane in its distribution and var. *borealis*, despite its name, seems to be common in southern bottomland hardwood forests.

Microscopic features (similar for both varieties)   Spores 9−12 × 3.5−4.5 µm. Pleurocystidia 37−45 × 6−12 µm. Cuticle of cap of tangled hyphae (a lattice).

Observations   The specific epithet means two colors. Var. *borealis* was first collected in northern Michigan hence the name which means northern. We suspect it may turn out to be more abundant in the South than the North.

73b *Boletus bicolor* var. *borealis*   One-half natural size

74 *Boletus erythropus* Pers. : Fr.
& *Boletus hypocarycinus* Sing.

Identification marks Both species have red or orange red tube mouths that stain blue when bruised, brown caps, and lack a reticulum on the stalks. In *B. erythropus* the cap is yellow to yellowish along the margin and the stalk is dull orange red whereas in *B. hypocarycinus* the cap is brown overall and the stalk is carmine red over the base. Spore size, as noted below, is another character that separates these species.

Edibility Dan Guravich reports he has eaten *B. erythropus* without ill effect, but members of this group of boletes are known to cause gastrointestinal upsets, and the odds are against the mycophagist.

When and where Both species fruit under hardwoods in the summer and early fall. *B. erythropus* is widely distributed in eastern North America; *B. hypocarycinus* was described from Florida and is shown here from Mississippi where it is abundant.

74a *Boletus erythropus*   About one-half natural size

74b *Boletus hypocarycinus*   About one-half natural size

Microscopic features   Spores of *B. erythropus* 12−15 × 5.5−6.5 μm; spores of *B. hypocarycinus* 9−11 × 3.5−4 μm.

Observations   *Erythropus* means red stalk; *hypocarycinus* seems to mean under nut-bearing trees.

75   *Boletus pseudosulphureus*   One-half natural size

Boletus pseudosulphureus Kallenbach   75

Identification marks   This bright yellow bolete quickly changes to deep blue or blackish blue when touched or injured. Neither the cap nor the stalk is slimy as in *B. curtisii*. The stalk is not reticulate, although exceptions occur as shown in the photograph. The cap is smooth, not powdery, and often streaked with dull red. No veil is present.

Edibility   We have no reliable information on it; not recommended.

When and where   Solitary to scattered on the ground in woods of various types, often under pines in the South where it fruits in late summer and fall; widely distributed in eastern North America and Europe but seldom common.

Microscopic features   Spores 13.5−15 × 4.5−5.5 μm. Pleurocystidia 45−65 × 8−12 μm. Cuticle of cap of tangled ascending hyphae 4−8 μm wide.

Observations   The specific epithet means false *sulphureus*, in this case a reference to another yellow bolete, *B. sulphureus*.

Boletus longicurvipes Snell & Sm.   76

Identification marks   This slender bolete combines traits of *Boletus* and *Leccinum*. The greenish blue discolorations in the base of the stalk of mature specimens and rough ornamentation on the stalk are reminiscent of

76 *Boletus longicurvipes*   Just under natural size

Leccinum; however, the ornamentation does not darken in age so we place it in *Boletus*. The bright yellowish brown cap is distinctly slimy.

Edibility Probably edible; it is closely related to *B. rubropunctus* which is edible.

When and where Solitary to scattered in the Appalachian highlands north to southern Michigan; summer and fall; it was first found in the Great Smoky Mountain National Park in Tennessee. We have collected it in North Carolina where white pine occurs with an understory of rhododendron.

Microscopic features Spores (11) 12−15 × 4.5−6 µm. Pleurocystidia 35−60 × 8−14 µm. Cuticle of cap an ixotrichodermium of narrow erect hyphae in a thick matrix of slime.

Observations *Boletus rubropunctus* (*Leccinum rubropunctum* of some authors) has brighter yellow tube mouths, larger spores (17−21 × 5−7 µm), broader end cells in the cuticle of the cap, and a thinner slime layer on the cap. It appears to be associated with oak and/or chestnut, and also occurs in the South. *Longicurvipes* refers to the long, curved stalk; *rubropunctus* means red dots.

77   *Boletus griseus* Frost in Pk.

Identification marks Pale gray to brownish gray tones predominate on these fruiting bodies including the tube mouths which are pale gray when young and gray brown at maturity. The stalk is distinctly reticulate and sometimes yellow near the base.

Edibility Edible and considered good; the raw flesh is mild.

When and where Widely distributed in eastern North America and sometimes abundant; generally associated with oak; summer and fall. In North Carolina we have found it from the western mountains to the Atlantic coast.

Microscopic features Spores 9−10.5 (13) × 3−4.5 µm. Pleurocystidia 30−60 × 6−10 µm, content dull brownish yellow as revived in KOH. Cuticle of cap a trichodermium that soon becomes a lattice.

Observations *B. griseus* and *B. retipes* seem to be the extremes in a series of boletes that vary in color from gray overall to yellow overall and whose flesh varies from mild and white to bitter and bright yellow. The mild, softer-fleshed (and usually wormier), gray extreme is edible whereas the bitter, hard-fleshed, yellow extreme is inedible because of its taste, if not actually poisonous. *Griseus* means gray.

77 *Boletus griseus*   About one-half natural size

78 *Boletus retipes*   About one-half natural size

Boletus retipes Berk. & Curt.   **78**

Identification marks Young caps are olive gray to yellowish brown with a chrome yellow margin and dry to moist, old ones are grayish to brownish yellow and often slightly viscid. Rusty brown stains develop on the chrome yellow tube mouths when they are injured. The stalk is bright yellow and ornamented with a conspicuous reticulum. The flesh is yellow, firm to hard, bitter, not bluing, and almost never wormy. One's hands are stained yellow after handling these boletes.

Edibility Unpalatable; the bitterness is not destroyed by cooking.

When and where Summer and fall in mixed or deciduous woods; often gregarious; described from Hillsborough, North Carolina, and widely distributed and common in the South, less common in the North.

Microscopic features Spores (9) 10.5–13.5 × 3.5–5 µm. Pleurocystidia 45–70 × 9–14 µm, contents yellowish brown in KOH. Cuticle of cap a lattice to cutis of narrow hyphae.

Observations *B. griseus* and *B. retipes* may be the extremes in a confusing series of boletes with varying amounts of gray and yellow color and bitterness. The specific epithet refers to the reticulate stalk. This species is also known as *Pulveroboletus retipes*.

---

79 *Boletus variipes* Pk.

Identification marks Cap color varies from deep grayish brown to rich yellow brown suggestive of buttered toast. Young tube mouths are plugged ("stuffed" in mycological parlance) by a weft of hyphae and are whitish, by maturity they become greenish yellow. A more or less distinct reticulum is present on the stout stalk. Young caps are dry to the touch. There are no color changes on injured areas.

Edibility Edible and good; discard the occasional bitter specimens. Insect larvae (the "worms" of wormy mushrooms) also like this bolete and only young buttons are usually free of them.

When and where Widely distributed in eastern North America, typically under hardwoods, occasionally with pine; late spring into fall. Only a few specimens are needed for a meal.

Microscopic features Spores 12–13.5 (16) × 3.5–4.5 µm. Pleurocystidia typically absent. Cuticle of cap a trichodermium.

Observations The specific epithet means variable stalk or foot. The steinpilze or cèpe, *B. edulis*, differs in having a slightly tacky to viscid cap and is usually associated with conifers.

79 *Boletus variipes*   One-half natural size

Boletus pinophilus Pilát & Dermek 80

Identification marks In young specimens the tube mouths are plugged or stuffed with a weft of white hyphae, later the tube mouths open and become yellowish green. The caps are deep brownish red to dull red and dry. A distinct reticulum is present on the stalk. No pronounced color changes occur on injured areas.

Edibility We can personally attest to the excellence of this species as an esculent; we ate part of the collection that is illustrated.

When and where Scattered to gregarious under pines in the coastal plain in summer; its exact range in North America is not known. The species was described from Czechoslovakia.

Microscopic features Spores 12–15 × 4–5 μm. Pleurocystidia not observed but reported to be 32–60 × 4.5–12 μm. Cuticle of cap of dry interwoven hyphae with ascending tips.

Observations *B. edulis*, a close relative of this species, has a yellow brown to dark brown cap that is thinly slimy in moist weather; it is also edible and good. The entire group of boletes in which the tube mouths are stuffed at first is in need of detailed study; there are about as many ideas on their classification as there are specialists in the group. *Pinophilus* means pine loving.

80 *Boletus pinophilus*   About one-third natural size

Boletus affinis Pk. 81

Identification marks Spore deposits are rusty gold to amber brown, bright for a bolete. The fruiting bodies are of medium size and have dry caps, white, unchanging flesh, and buff to yellowish tan tube mouths. The stalk is tinged with the color of the cap and at most only faintly reticulate at the apex. In var. *maculosus* the cap is deep yellow brown with scattered

81 *Boletus affinis*   About two-thirds natural size

lighter spots. In var. *affinis* it varies from purplish brown (brighter than pecan shells but similar) to dull yellowish brown and is unspotted.

Edibility   Edible but seldom found in quantity.

When and where   Described from New York and widely distributed in eastern North America under hardwoods; summer and fall.

Microscopic features   Spores 10−14 × 3−5 μm. Pleurocystidia (30) 45−55 × 6−9 (12) μm, contents often rusty orange in KOH. Cuticle of cap a trichodermium; lower cells barrel-shaped to subglobose; end cells predominantly clavate.

Observations   The specific epithet means related; *maculosus* means spotted. This species is called *Xanthoconium affine* by some authors. *Xanthoconium* is distinguished from *Boletus* by the lack of olive tones in the spore deposit.

---

82   *Boletus viridiflavus* Cok. & Beers

Identification marks   Among the species of *Boletus* with brilliant yellow tube mouths (at least when young), this one can be distinguished by the combination of olive gold cap which may have reddish tints and is viscid when wet, red to rose stalk, and greenish yellow hymenophore in age. Abundant white mycelium occurs at the base of the stalk. The fruiting bodies are usually slender. Tube mouths of fresh specimens are reported to turn dull brick red when bruised.

Edibility   We have no reports on it; not recommended.

When and where   Scattered to gregarious under hardwoods, in mixed woods, and in open areas; summer through fall; widely distributed in the Southeast west into Texas. The type collection was gathered at Highlands, North Carolina.

Microscopic features   Spores (12) 13.5−16 × 4.5−6 μm. Pleurocystidia 45−63 × 9−12 μm. Cuticle of cap an ixolattice in which, apparently, the walls of the hyphae gelatinize.

Observations   The specific epithet is formed from words meaning green and yellow.

82 *Boletus viridiflavus*   One-half natural size

Boletus caespitosus Pk.   83

Identification marks   The rich brown of the cap is dull in comparison to the persistently brilliant yellow tube mouths. The caps are velvety and often dull purplish red near the margin. No color changes occur on bruised tubes or flesh. The fruiting bodies usually occur in clusters, i.e., they are cespitose as indicated by the specific epithet.

Edibility   We have no reliable information on it; not recommended.

When and where   Summer and fall under hardwoods; first collected in Virginia and reported from Nova Scotia and North Carolina but probably more widely distributed.

Microscopic features   Spores 9–10.5 × 3.7–4.5 μm. Pleurocystidia 45–64 × 9–12 μm. Cuticle of cap a cutis.

Observations   The name *Pulveroboletus caespitosus* is sometimes used for this species. Some mycologists maintain *B. caespitosus* and *B. auriporus* as distinct species, others combine them into *Boletus* or *Pulveroboletus auriporus*. Clearly there are problems yet to be solved in this group.

83 *Boletus caespitosus*   About one-half natural size

84 *Boletus fraternus* Pk.

Identification marks By maturity the surface of the cap is cracked, exposing the yellow context between patches of intact cuticle. The caps are deep red at first, dull red in age, dry, and seldom over 6 cm broad. The context is yellow and turns blue when injured as do the large, angular tube mouths.

Edibility Not recommended because of lack of information and the difficulty of making correct identifications in this group.

When and where Gregarious to cespitose in thin woods, especially under hardwoods, in wooded parks, lawns, etc., along roadsides, and even on stumps; summer and early fall; common and widely distributed in the Southeast. Underwood found the type collection in July along the shaded streets of Auburn, Alabama.

Microscopic features Spores 10.5−12 (16) × 4.5−6 µm. Pleurocystidia 30−55 × 5−8 µm. Cuticle of cap a trichodermium that soon becomes separated into patches by expansion of the cap.

Observations This species is in the same group as *B. campestris*; none of these have a reputation for causing serious poisonings; neither are they particularly good eating. *Fraternus* means brotherly or closely allied.

84 *Boletus fraternus*   About one-half natural size

85 *Boletus campestris* Sm. & Thrs.

Identification marks Several species of bolete share these features: relatively small fruiting bodies; dry, red, and often velvety caps; yellow tube mouths that stain blue when injured; and yellow, pruinose to glabrous stalks. They are distinguished on spore size, pore size, and structure of the cuticle of the cap—all characters difficult to deal with in the field.

Edibility Not recommended; see comments on *B. fraternus*.

When and where Members of this group are to be expected in open woods, shaded lawns, and parks, and similar areas usually under hardwoods; they fruit in the summer and fall and are, as a group, common and widely distributed in eastern North America.

Microscopic features Spores 10.5–13.5 × 4.5–6 μm. Pleurocystidia 37–55 × 7.5–11 μm. Cuticle of cap initially a trichodermium, in age of interwoven ascending hyphae, amyloid granules present in some cells, end cells tapering and walls roughened.

Observations *B. campestris* is a common suburban bolete in western Mississippi and in Michigan. The specific epithet means of the fields. Accurate identifications in this group can only be made by study of microscopic features.

85 *Boletus campestris*   About one-half natural size

86 *Phylloporus rhodoxanthus*   One-half natural size

Phylloporus rhodoxanthus (Schw.) Bresadola   **86**
(Gilled Bolete)

Identification marks Minutely velvety to suedelike yellow brown to reddish brown caps; decurrent, distant, dull yellow to ochraceous gills; soft flesh; and relatively smooth stalks are characteristic of this species. The surface of the cap turns bright blue to deep bluish green when a drop of household ammonia is placed on it. Both the odor and taste are mild. This is a collective species, some specimens stain blue when injured, others do not; the mycelium at the base of the stalk may be yellow or white; and the gills may be yellow and distinct, or dull tan, interconnected, and somewhat boletinoid.

Edibility Edible but inclined to be soft when cooked; opinions vary on its desirability.

When and where Solitary to scattered in woods of various kinds, widely distributed in the forested regions of North America but seldom abundant; late spring to late fall in the South.

Microscopic features Spores 9–10.5 (12) × 4–5 µm. Pleurocystidia 37–52 (111) × 10–15 (18) µm. Cuticle of cap a trichodermium.

Observations In all respects except the configuration of the spore-bearing tissue this species is a bolete and is so classified in most current works. The genus name is derived from words that mean gill and pore; the specific epithet from ones meaning rose and yellow.

Russulaceae

Within the gilled mushrooms the diagnostic combination of characters for this family is the presence of clusters of globose cells (sphaerocysts) in the flesh of the fruiting bodies and spore ornamentation that turns black to bluish black (i.e., is amyloid) in iodine-containing solutions (see Melzer's reagent, p. 268). Of more help to the mushroom hunter is the absence of veils in North American species and the stout relatively fragile fruiting bodies that tend to shatter when thrown against a hard surface. With a little practice and study one can recognize members of this family without recourse to the microscopic characters. A few species in other families, e.g., *Amanita farinosa* can be mistaken for species in the Russulaceae if one is not careful.

Key to Genera

1. Broken or cut portions of the fruiting body, especially the gills, exuding a clear and colorless, opaque and white, or variously colored liquid (latex) . . . . (p. 114) *Lactarius*
1. Broken or cut fruiting bodies not yielding a liquid (if there is a color change on exposed flesh to green, see *Lactarius*) . . . . (p. 131) *Russula*

Lactarius

Among the gilled mushrooms this is the only genus in which all the species are capable of exuding a liquid, the latex, when the fruiting body is broken or cut. The gills of young specimens are often the best places to check for the presence of a latex. In our species there are no true veils, the gills are attached, and the fruiting bodies are often stocky.

Several edible species belong to this genus. Species with a strongly acrid (burning or peppery) latex and/or those in which the latex soon turns yellow or purple when exposed to air are not recommended for eating. Common names for this genus include milk mushroom and milk cap and, like the scientific name, refer to the milk or latex.

Key to Species

1. Latex colored (blue, orange, purple, muddy reddish brown, or dingy yellow) when first exposed to air 2

87   *Lactarius indigo* (Schw.) Fr.
(Blue Lactarius)

Identification marks   All parts of the fruiting body including the context and latex are blue at first; the cap often fades to grayish or silvery blue in age. Green stains develop in age and on injured areas. Fresh caps may be slightly tacky to slippery.

Edibility   Edible; this is a good species for beginners as it is the only all blue *Lactarius* and therefore easy to identify.

When and where   Scattered to gregarious in mixed woods, especially under or near oak; summer and fall or early winter when rain is plentiful. Schweinitz collected it in North Carolina; the species is widely distributed east of the Great Plains but most abundant in the South.

Microscopic features   Spores $6-7.5 \times 5.5-6.5$ µm, ornamented with a broken reticulum to 0.5 µm high. Pleurocystidia absent. Cuticle of cap an ixocutis.

Observations   Var. *indigo*, shown here, is the common variety and the best for eating. Var. *dimunutivus* has smaller, slenderer fruiting bodies; distant and somewhat decurrent gills, and fruits in damp areas under pine or in mixed woods; it has been reported only from Texas and North Carolina.

87 *Lactarius indigo*   One-half natural size

88 *Lactarius paradoxus*   Two-thirds natural size

Lactarius paradoxus Beards. & Burl.   88

Identification marks   Young caps are often silvery blue to greenish blue with a purple margin, old ones may be silvery green or a mixture of all three colors. The gills are close and purplish red to dull purple. The latex is brownish purple. Green stains develop in age and on injured areas of the fruiting body.

Edibility   Edible.

When and where   Solitary to gregarious under 2- and 3-needle pines, often abundant and conspicuous in shaded lawns in late fall. It was described from Florida and is widely distributed in eastern North America. This is one of several species that fruit in late summer and early fall in the North and during fall and early winter in the South.

Microscopic features   Spores 7.5–9 × 6–7 µm, ornamented with a broken to partial reticulum to 0.7 µm high. Pleurocystidia absent to rare, 42–50 × 2–5 µm. Cuticle of cap an ixocutis.

Observations   The specific epithet means paradoxical, i.e., apparently self-contradictory or incredible. *L. subpurpureus* also has purplish colors on the fruiting bodies but has distant gills, lacks tones of blue, and is found where hemlocks grow in the mountains and farther north.

Lactarius salmoneus Pk.   89

Identification marks   The contrast of the white cap with the orange gills, stalk, flesh, and latex is at once striking and distinctive. Green stains develop on some specimens. The relatively small size of the fruiting bodies is also characteristic.

Edibility   Presumably edible but the small size of the fruiting bodies is a disadvantage.

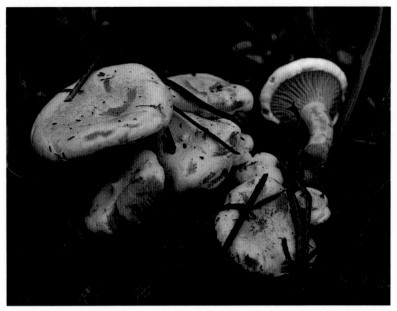

89 *Lactarius salmoneus*   About natural size

When and where   Occasionally solitary, typically densely gregarious under pines in damp areas or in wet weather from summer until near the end of the growing season. Peck described the species from specimens gathered near Auburn, Alabama. It is widely distributed in the South but apparently fruits sporadically.

Microscopic features   Spores 6.5–8 × 5–6 µm, ornamented with a broken to partial reticulum to 0.4 µm high. Pleurocystidia when present 34–46 × 6–8 µm. Cuticle of cap of dry interwoven and ascending hyphae.

Observations   *L. deliciosus* is another species with orange milk but is easily recognized by the combination of bright to dull orange cap that is thinly viscid in wet weather and is usually at least twice the size of those of *L. salmoneus. L. deliciosus* is edible and popular with many people. *Salmoneus* presumably refers to the salmon orange latex.

---

90   *Lactarius pseudodeliciosus* Beards. & Burl.

Identification marks   Injured areas slowly stain bluish green; however, the scanty latex is dull orange to rusty orange when first released. The cap is dull cream color to buff or light ochraceous as are the gills and stalk. The taste is peppery to acrid.

Edibility   Presumably edible; no species of *Lactarius* that stains green is known to be poisonous.

When and where   Late fall and winter under oak and pine; described from Florida and to be expected in the southern coastal plain generally.

Microscopic features   Spores 7–9 × 6–7 µm, ornamented with a broken to nearly complete reticulum. Pleurocystidia inconspicuous, 44–53 (80) × 5–8 µm. Cuticle of cap a thin ixocutis.

Observations   The specific epithet means false *deliciosus*, a reference to its relationship with *L. deliciosus*. The latex is often so scanty that one might think it is absent, but the bluish green stains are a signal that one has a *Lactarius* rather than a *Russula*.

90 *Lactarius pseudodeliciosus*   About one-third natural size

Lactarius atroviridis Pk.   **91**

Identification marks   This is the only common *Lactarius* in North America with a dark green cap and stalk. The gills are pinkish tan when young and stained brown in age. Darker green spots frequently are present on the cap and stalk. The latex is white and acrid. When touched with a drop of potassium hydroxide or ammonium hydroxide, the green parts turn magenta.

Edibility   Not recommended.

When and where   Typically gregarious under or near oaks; summer and fall; described from eastern New York and widely distributed in eastern North America, abundant at times in the southern Appalachians.

Microscopic features   Spores 7−9 × 5.5−6.5 µm, ornamented with a broken to partial reticulum. Pleurocystidia 45−60 (120) × 6−12 µm, abundant to absent. Cuticle of cap of appressed hyphae.

Observations   These are hard fungi to find because of their squat stature and dull coloration. The specific epithet means blackish green.

91 *Lactarius atroviridis*   One-half natural size

92 *Lactarius luteolus* Pk.

Identification marks Important features of this species are the velvety, whitish to pale creamy yellow cap and stalk; medium to small size of the fruiting bodies; and copious mild latex that stains injured areas brown. It has a rather strong fetid to fishy odor. Unlike several whitish species of *Lactarius* of this general size, no pink, purple, or reddish stains develop.

Edibility Edible.

When and where Scattered to gregarious under hardwoods and in mixed deciduous and coniferous woods east of the Great Plains. It fruits in the summer and fall, as late as November in the South and as early as June in the North.

Microscopic features Spores 7–9 × 5.5–7 µm, ornamented with warts and ridges to 0.8 µm high. Pleurocystidia 47–70 × 3–6 µm. Cuticle of cap a cellular zone bearing a dense turf of cystidia.

Observations The copious brown-staining latex and fishy odor remind one of *L. volemus* and *L. corrugis* but the color is quite different and the cystidia are thin walled. *Luteolus* means yellowish.

92 *Lactarius luteolus*   Just under natural size

93 *Lactarius corrugis* Pk.

Identification marks Wrinkles (particularly near the margin) on light to dark reddish brown to orange brown caps; pale gills that stain brown where injured; and copious, mild to astringent latex distinguish this species. The cap is dry and velvety; the stalk is similar in color to the cap or lighter.

Edibility Edible and good; the copious sticky latex interferes with frying; we like it baked in a cream sauce with bacon and onions.

When and where Solitary or more often gregarious under hardwoods and in mixed oak-pine woods; summer and early fall. It was described from New York and is widely distributed in eastern North America. We have seen large fruitings of it in the Piedmont and southern Appalachians.

93 *Lactarius corrugis*   One-half natural size

Microscopic features Spores 9–12 μm in diameter; ornamentation a complete reticulum 0.2–0.7 μm high. Pleurocystidia 60–125 (200) × 6–10 (15) μm, walls thickened to 7 μm. Cuticle of cap a turf of thick-walled cells arising from a cellular zone.

Observations The specific epithet refers to the wrinkles or corrugations on the cap. Compared to *L. volemus*, this species typically has darker, more wrinkled caps, stouter fruiting bodies, and larger spores.

Lactarius volemus (Fr. : Fr.) Fr.   **94**

Identification marks Relatively smooth caps; whitish to cream-colored gills that stain brown when injured; and the copious, mild, white to cream-colored latex are important characters. The latex stains injured areas brown and often smells fishy. The cap and stalk are tawny to orange yellow in var. *volemus* and whitish to clear yellow in var. *flavus*.

Edibility Edible, good in casseroles (see *L. corrugis*).

When and where Scattered to gregarious in deciduous and pine-oak woods and shaded lawns with oaks; widely distributed east of the Great Plains. It often fruits in large quantities in the South and is most abundant during hot wet weather in the summer and early fall.

Microscopic features Spores 7.5–9 μm in diameter, ornamented with a complete reticulum to 0.8 μm high. Pleurocysitida 48–145 × 5–13 μm, wall to 3 μm thick. Pilear cuticle a zone of large cells surmounted by a turf of cystidia.

Observations *Volemus*, according to our dictionary, refers to a kind of pear. Compared with *L. corrugis*, the fruiting bodies are slenderer, paler, smoother on the caps, and the spores are smaller.

94 *Lactarius volemus*   About one-half natural size

95   **Lactarius piperatus** (L. : Fr.) Pers.
(Peppery Lactarius)

Identification marks   White, medium to large fruiting bodies; extremely crowded narrow gills; exceedingly acrid flesh and latex; and glabrous cap and stalk are characteristic of this species. Dried droplets of latex on the gills are creamy yellow in var. *piperatus*, grayish green in var. *glaucescens*.

Edibility   None of the large, white, and acrid species of *Lactarius* can be recommended for eating; however, some people do boil these mushrooms in several changes of water and eat what is left. This species has caused severe gastrointestinal upsets.

When and where   Solitary to gregarious in deciduous and mixed woods containing oak; summer and early fall; widely distributed east of the Great Plains and often abundant.

Microscopic features   Spores 5−7.5 × 5−6.5 μm, ornamented with low warts and ridges to 0.2 μm high. Pleurocystidia 35−70 × 5−9 μm. Cuticle of cap of inflated cells and scattered cystidia.

Observations   *Piperatus* refers to the peppery to acrid taste of the raw flesh.

95 *Lactarius piperatus*   One-third natural size

Identification marks   In young specimens the margin of the cap is in-rolled and both thick and soft (almost cottony); in age the margin is thin. The latex is white and acrid—the longer one tastes it, the stronger the flavor. At first the entire fruiting body is white but as it matures and ages it becomes tan to pale brown and brown stains develop on the gills. The stalk is velvety (use a hand lens).

Edibility   Untested but not recommended because of the acrid taste.

When and where   Scattered in woods, often in dry pine woods or under hardwoods; summer and early fall; widely distributed in eastern North America but the details of its distribution remain to be learned.

Microscopic features   Spores 7.5–9 (11) × 6–7 (8.5) μm, ornamented with isolated warts to 1 μm high. Pleurocystidia 57–75 (90) × 6–9 μm. Cuticle of cap of loosely interwoven hyphae.

Observations   The specific epithet refers to the tomentose margin of the young caps. This species is closely related to *L. deceptivus* which has larger fruiting bodies, larger spores (9–12 × 7.5–9 μm) with higher ornamentation (to 1.5 μm tall), and a penchant for damp mixed woods especially where hemlock is present.

96   *Lactarius tomentoso-marginatus*   About one-third natural size

Lactarius allardii Cok.   **97**

Identification marks   Young caps and those covered by dirt or leaves are white, in age and/or where exposed to light, the caps become light to dark cinnamon to pale brick red. The acrid latex is white at first and dries grayish green. Frequently the caps are lopsided rather than circular in outline. Fruiting bodies of this species are among the largest in the genus.

Edibility   Not recommended because of the strongly acrid taste.

When and where On soil in deciduous or mixed woods, especially those containing oak and pine. It fruits in the summer and early fall. This is primarily a southern species and was described from Chapel Hill, North Carolina. It is also common in the sandy oak-aspen-pine forests of northern Michigan in late summer.

Microscopic features Spores 7.5–10.5 × 6–8 µm, ornamented with low warts and fine lines, not reticulate. Pleurocystidia 60–110 × 7–14 µm. Cuticle of cap a cutis.

Observations Coker named this species for his first assistant, A. H. Allard.

97 *Lactarius allardii*   One-third natural size

98 *Lactarius croceus*   Two-thirds natural size

Lactarius croceus Burl.   98

Identification marks   The brilliant orange yellow color of all parts of the fruiting body, latex which is white at first and quickly changes to yellow orange, and the cap that is slimy at first form a unique combination of characters. The flesh is bitter to acrid. No fascicles or tufts of hyphae occur on the cap margin.

Edibility   We have no information on it but do *not* recommend eating any *Lactarius* with a strongly peppery taste and/or whose latex turns yellow when exposed to the air.

When and where   Summer and early fall in deciduous woods. Burlingham collected it in North Carolina, and it occurs from the southern Appalachians north to southern New England and Ohio.

Microscopic features   Spores 7−9 (10.5) × 6−7 μm, ornamented with a broken reticulum 0.3−0.6 μm high. Pleurocystidia 37−60 (75) × 5−9 μm. Cuticle of cap a thick ixocutis.

Observations   Among the species of *Lactarius* whose latex turns yellow and that have a viscid cap, this is the only common orange species. *Croceus* means saffron colored.

Lactarius chrysorheus Fr.   99

Identification marks   The cap is pale pinkish orange, pale yellowish cinnamon, or dull orange tan, zonate, and moist to subviscid. The latex quickly changes from white to sulphur yellow on cut surfaces of the fruiting bodies. The stalk is dry, whitish, and does not darken appreciably in age. Purplish red stains do not develop in age or on injured areas.

Edibility   Often reported to be poisonous, hence not recommended. The taste is strongly acrid.

99   *Lactarius chrysorheus*   Two-thirds natural size

When and where   Common from late summer into early winter on sandy soil where oaks grow; common in eastern North America and reported from the Pacific Northwest.

Microscopic features   Spores 7–8 × 5–6 µm, ornamentation to 1 µm high, of warts and ridges forming at most a broken reticulum. Pleurocystidia 50–80 × 6–11 µm. Cuticle of cap a thin ixocutis.

Observations   The specific epithet means exuding a golden liquid. *L. vinaceorufescens* has often been confused with this species. In it the caps are pinkish tan, typically without spots and zones, and purplish red stains develop on old specimens, particularly at the base of the stalk. It is typically associated with pines and both species may be found in pine-oak woods.

---

100   *Lactarius subvernalis* Hes. & Sm.

Identification marks   The close to crowded gills and whitish to pale grayish tan coloration are characteristic. Furthermore, the latex is white at first and injured areas stain salmon to dull light purple. The taste is quite acrid.

Edibility   Not recommended because of the acrid taste but we have no data on its toxicity.

When and where   Scattered to gregarious under hardwoods and in mixed woods; late spring into fall. The extremes of its known distribution are Massachusetts and Texas but it is widely distributed and common in the South. The type collection was gathered near Coldspring, Texas.

Microscopic features   Spores 6–8 × 6–7 µm, ornamentation to 1.5 µm high as a broken to partial reticulum. Pleurocystidia limited to region near gill edge and similar to cheilocystidia, (37) 55–72 × (3) 5–7 µm. Cuticle of cap of ascending hyphae with inflated, clavate to pyriform cells.

Observations   It often fruits at the same time and in the same habitat as *L. subplinthogalus* but the difference in gill spacing is an easy way to distinguish them. The specific epithet refers to the time of year the type collection was gathered (May), nearly spring. The prominent cystidia are unusual in the group of milk mushrooms to which this one belongs.

100   *Lactarius subvernalis*   Two-thirds natural size

101 *Lactarius subplinthogalus*   About two-thirds natural size

Lactarius subplinthogalus Cok.   **101**

Identification marks The distant (widely spaced) gills, pale coloration, and dull reddish salmon to rusty orange stains that slowly develop on injured areas are important. The cap is smooth but not velvety; it and the stalk are very pale pinkish tan to grayish tan when young and slightly darker in age. The latex is white when first exposed and slowly changes color; it is distinctly acrid.

Edibility Not recommended because of the acrid taste.

When and where Solitary to scattered under hardwoods and in mixed woods, to be expected throughout the South but most abundant outside the high mountains; summer and early fall; described from North Carolina.

Microscopic features Spores 6.5–8 (9) μm in diameter, nearly globose; ornamentation to 2.5 μm high, forming a broken reticulum. Pleurocystidia not observed. Cuticle of cap of ascending, short-celled hyphae with clavate to subcylindric end cells.

Observations In some collections there appears to be an extremely thin layer of slime on the cap; however, they do not feel slippery. The specific epithet means near *L. plinthogalus*.

Lactarius peckii Burl.   **102**

Identification marks Important features of this species include the orange brown to rusty brown or terra-cotta, zoned cap; dull reddish brown to dull purple brown close to crowded gills; and white, copious latex. The latex dries dull pale gray green or pale yellow. The gills are among the darkest in the genus.

Edibility Not recommended because of the strongly acrid taste.

102 *Lactarius peckii*   About two-thirds natural size

When and where   July into September in well-drained woods where oak is present, common in the South and known as far north as Rhode Island. Burlingham collected it in the Pink Beds of North Carolina.

Microscopic features   Spores 6–7.5 μm in diameter, ornamented with a heavy partial to complete reticulum to 0.8 μm high. Pleurocystidia 37–52 × 4.5–7 μm. Cuticle of cap of dry hyphae forming a turf when young, becoming interwoven in age.

Observations   C. H. Peck, for many years state botanist of New York, is commemorated in the specific epithet. Peck has been called "the father of systematic American mycology."

---

103   *Lactarius argillaceifolius* Hes. & Sm.

Identification marks   The cap is grayish tan to pinkish gray with a paler margin. It is slimy in wet weather and may be glazed and dry at other times. The stalk varies from greasy to dry depending on the weather and age of the fruiting body. The dull buff latex dries light tan to dull olive. In age the gills become dull yellowish brown, clay color to mycologists.

Edibility   This species is one of several lumped together for many years under the name of *L. trivialis* and no accurate data on its edibility is available. The taste is acrid in young specimens which detracts from its merits as an esculent.

When and where   Solitary to scattered under hardwoods and in mixed woods where oak grows; midsummer to fall; sporadically abundant.

Microscopic features   Spores 7.5–9 × 6–8 μm, ornamentation to 0.5 μm high, forming at most a broken reticulum. Pleurocystidia 40–80 × 8–12 μm. Cuticle of cap varying from an ixotrichodermium to an ixocutis.

Observations   The specific epithet means clay colored (dingy yellowish brown) gills. *L. fumeacolor*, known from Florida, is similar in appearance but has larger spores (9–11 × 7–8 μm).

Identification marks   Orange to orange tan caps that have tufts of hyphae embedded in slime and are faintly zoned are characteristic of this species. When young the margin is fringed with clumps of hyphae that are less obvious in age. The acrid latex is white when first exposed to the air and dries the color of cream on white paper. Rusty brown stains may develop on the gills in age. The odor is faintly fragrant to pumpkinlike.

Edibility   Not recommended because of the acrid taste.

When and where   Gregarious to scattered under hardwoods or in mixed woods; described from the Pink Beds of western North Carolina, and to be expected in the southern Appalachians; August and September.

Microscopic features   Spores 7–8 × 5–6 µm; ornamentation to 1 µm high, in the form of a broken reticulum. Pleurocystidia 37–60 × 7–11 µm. Cuticle of cap an ixocutis with embedded tufts of hyphae.

Observations   We take pleasure in illustrating this little-known species. The specific epithet means to adhere or unite with glue.

103   *Lactarius argillaceifolius*   Two-thirds natural size

104   *Lactarius agglutinatus*   Three-fourths natural size

105 *Lactarius yazooensis* Hes. & Sm.

Identification marks Concentric bands of pale orange and yellow orange are present on the cap which is sticky in wet weather and is dry and appears glazed in dry weather. There are no hairs or tufts of hyphae on the margin of the cap at any age. The stalk is short, unspotted, and hard. The latex is white, unchanging, and extremely acrid.

Edibility Not recommended; we have no data on it but the extremely acrid taste makes it unappetizing at the very least.

When and where Gregarious to scattered in lawns and grassy places where oaks grow and in thin woods; summer and fall; to be expected generally in the coastal plain and Florida but known only from Mississippi to date.

Microscopic features Spores 7–9 × 6–7.5 µm, ornamented with short ridges and warts which form a pattern resembling the markings of a zebra. Pleurocystidia 37–60 × 4.5–8 µm. Cuticle of cap a thin ixocutis.

Observations This species was first collected by Dan Guravich and is named for the Yazoo River of Mississippi. It is one of a group of closely related species that differ primarily in microscopic features. *L. psammicola*, another member of the group, has a hairy cap margin at first and deeper orange colors on the cap. It is also common in the South but is too acrid to be attractive to the mycophagist.

105 *Lactarius yazooensis*　　　　　　　　　　About three-fourths natural size

106 *Lactarius hygrophoroides* Berk. & Curt.

Identification marks The caps are consistently dry, velvety, and often slightly wrinkled. They vary from orange cinnamon to rusty orange brown or salmon orange in color. The gills are distant in contrast to those of *L. volemus* and *L. corrugis*. The latex is white, mild, and in some varieties dries pinkish gray.

Edibility Edible and good.

When and where Widely distributed east of the Great Plains in deciduous or mixed woods; late spring to early fall. The largest fruitings we have seen were in Florida in August where one could have easily picked a peck of specimens in a few minutes.

106 *Lactarius hygrophoroides*   Natural size

Microscopic features Spores 7.5–9.5 (10.5) × 6–7.5 µm; ornamentation fine, of warts and lines 0.2–0.4 µm high. Pleurocystidia absent. Cuticle of cap a zone of inflated cells with some projecting cystidia.

Observations As usual the more a group is studied, the more variations come to light, so it is no surprise that there are several varieties of this species that differ in taste and the color of the dried latex. The distant gills, smaller stature, and lack of brown stains separate this species from *L. volemus* and *L. corrugis* in the field. *Hygrophoroides* means *Hygrophorus*-like.

Russula

Most species of *Russula* have squat and/or stout fruiting bodies that can easily be recognized as belonging to this genus. The stalk is typically equal in diameter throughout its length, never has an annulus or other sign of a veil, and is often white or light in color; it either easily crumbles when crushed, or, in a few species, is quite hard. The spores vary from white to pale yellow to dark ochraceous in deposits depending on the species and all have amyloid ornamentation. No latex is exuded from cut surfaces. In most species the cap is glabrous and brightly colored.

This is a large genus and one that is still in need of considerable study before many of the North American species will be easy to identify. Although we counsel against eating any mushroom whose identity cannot be accurately established, many people do eat those species of *Russula* that have a mild flavor. Species whose fruiting bodies stain red, dark brown, black, or a combination of these colors, should, however, be avoided.

Key to Species

1. Cap green at maturity . . . . 107. *Russula aeruginea*
1. Cap some other color at maturity or white 2
   2. Cap white when young and at maturity . . . . . 108. *R. romagnesiana*
   2. Not as above 3
3. Cap white to buff when young, soon becoming pale brown, tawny, or cinnamon; injured areas staining light cinnamon brown to tawny
   . . . . 109. *R. compacta*

3. Not as above 4
    4. Cap dry, dark grayish brown to brownish gray . . . .
        . . . . 110. *R. subnigricans*
    4. Not as above 5
5. Brick red to brownish red crustose patches present on the cap and
    base of the stalk . . . . 111. *R. balloui*
5. Not as above 6
    6. Cap bright rusty orange to rusty yellow; stalk and gills slowly
        staining yellow when injured . . . . 112. *R. subfoetens*
    6. Not as above 7
7. Cap at maturity darker grayish brown over the center than near the
    margin . . . . 113. *R. amoenolens*
7. Cap more or less evenly colored, pale orange yellow, yellowish tan,
    light yellowish brown, etc. 114. *R. pectinatoides*

---

107 *Russula aeruginea* Lindblad in Fr.

Identification marks The cap is grayish olive green when young and grayish yellow green to yellowish green at maturity. The gills and stalk are white to ivory and the spores are pale orange yellow in deposits. Both the odor and taste are basically mild.

Edibility Edible.

When and where Scattered to gregarious under hardwoods and in mixed woods; summer and fall; widely distributed in eastern North America and common.

Microscopic features Spores 6–8 (9) × (5) 6–7.5 μm, ornamented with warts, some connected by low ridges but not forming a reticulum. Cuticle of cap an ixocutis giving rise to scattered clusters of pileocystidia.

Observations *Aerugineus* refers to the color of verdigris, i.e., green or greenish blue. This is not the only species of *Russula* with a green cap in the South, but they are difficult to separate without reference to microscopic characters. There are no reports that any are poisonous, as far as we know.

107 *Russula aeruginea*   About one-half natural size

108 *Russula romagnesiana*   One-half natural size

Russula romagnesiana Shaffer   **108**

Identification marks Both the cap and stalk are creamy white when young and fresh but stain dull yellow to brown in age and where injured. The gills are close and pale olive buff; short gills (lamellulae) are present. The taste is mild to slightly acrid. The spores are white in deposits.

Edibility Unknown, not recommended.

When and where Only a few collections of this species have been made, some in Michigan and this one in Mississippi, its distribution remains to be ascertained. It fruits in the summer and early fall under hardwoods, especially oak as well as under pine.

Microscopic features Spores 6–8 × 6–7 μm (excluding ornamentation); ornamentation to 1.5 μm high as blunt to acute warts sometimes joined by lines. Cuticle of cap of matted loosely interwoven hyphae.

Observations The species is named for Henri Romagnesi, the well-known French mycologist who has studied the genus *Russula* intensively. This species has no doubt been confused with *R. brevipes* in the field, but the latter has spores 8–10.5 × 6.5–10 μm that are somewhat reticulate; the fruiting bodies are similar in appearance. We include *R. romagnesiana* to emphasize that identification made on field characters must be confirmed by a study of microscopic features before being considered final.

Russula compacta Frost in Pk.   **109**

Identification marks Young caps may be tacky and are white to pale buff or yellowish; as they age they become dry and darken to cinnamon buff, pale brown, cinnamon, or tawny. The margin is not striate. The gills stain

109 *Russula compacta*   One-half natural size

light brown where injured and often fork near the stalk. The stalk is hard at first and colored like the cap. Old specimens have a strong, fishy, unpleasant odor but the taste is mild to slightly bitter. The spores are white in deposits.

Edibility   Reported to be edible but if the flavor of cooked specimens resembles the odor, not likely to be popular.

When and where   On the ground under hardwoods and in mixed woods; as early as May along the Gulf coast through summer and fall at higher elevations and further north; not uncommon in eastern North America.

Microscopic features   Spores 7.5–9 × 6–8.5 μm (excluding ornamentation); ornamentation to 1 μm high of warts and ridges forming a broken to partial reticulum. Cuticle of cap a thick cutis of narrow hyphae.

Observations   The specific epithet means compact.

110   *Russula subnigricans* Hongo

Identification marks   Young specimens as shown in the photograph have a dry gray brown cap, pale gray to whitish stalk, and cream-colored to slightly pink gills. The gills are thick and subdistant. Injured areas on the fruiting bodies stain reddish to purplish lavender. In older, water-soaked specimens both the cap and the gills may be quite reddish lavender and then the fruiting bodies may remind one of some species of *Laccaria* or *Lactarius*.

Edibility   This is reported to be the only deadly poisonous *Russula* in Japan. We have no data on North American material but recommend avoiding this and other species of *Russula* whose fruiting bodies stain red, dark brown, black, or a combination of these colors.

When and where   Scattered under hardwoods; summer. Dan Guravich has found it in Mississippi several times, but otherwise its distribution in North America remains to be learned. It was described from Japan.

Microscopic features   Spores 6.5–9 × 6–7.5 μm, ornamented with low (less than 0.2 μm high) warts connected by fine lines. Cuticle of cap a dry cutis.

Observations Our identification of these specimens as *R. subnigricans* is still somewhat tentative, but the American material seems to fit the description of the Japanese species. *R. subnigricans* is poisonous and we include mention of it as a warning to American mycophagists that there may be a dangerously poisonous *Russula* in North America. *Subnigricans* means somewhat blackish or it could be interpreted to mean near *R. nigricans*.

110 *Russula subnigricans*   About one-half natural size

Russula balloui Pk.   **111**

Identification marks As the caps expand, the surface breaks up into small, flat, brick red to reddish brown, scablike scales on a dull yellow background. The caps are small, only 3−4 cm broad. The gills are pale yellow to white and do not fork. At first the surface of the stalk, especially near the base, is evenly light brick red to reddish brown but that layer soon breaks up into small scales.

Edibility Unknown, not recommended.

When and where Scattered under hardwoods; fall; uncommon to rare. The type collection was gathered on Staten Island, New York; we have seen it in Ohio and Mississippi but details of its distribution are yet to be learned.

Microscopic features Spores 7.5−9.5 × 7−8 µm, ornamentation to 1 µm high, forming at most a broken reticulum. Cuticle of cap an ixocutis to ixolattice with matted to erect rusty brown superficial hyphae.

Observations William Hosea Ballou (1857−1937) was a man of many interests, one of which was mycology. In 1908 he is reported to have founded a national movement to conserve wild mushrooms.

111 *Russula balloui*   Just under natural size

112   *Russula subfoetens* W. G. Smith

Identification marks   The cap is viscid, bright rusty orange to rusty yellow, and strongly convex to cushion-shaped when young but plane in age. The odor is distinct but weak to moderately strong and fetid to somewhat fragrant. The taste, particularly of the gills, is acrid. In age the margin of the cap is striate.

Edibility   Not recommended but not known to be poisonous.

When and where   Summer and early fall in thin woods and shaded lawns where hardwoods are present; widely distributed in eastern North America and abundant along the Gulf coast.

Microscopic features   Spores 7–8 × 5.5–6.5 μm; ornamentation to 1.5 μm tall, of isolated warts and ridges at most forming a broken reticulum. Cuticle of cap a thick ixocutis with patches of pileocystidia.

Observations   The stalk and gills slowly become yellow when injured. The specific epithet means somewhat fetid or bad smelling.

112   *Russula subfoetens*   About one-half natural size

113   *Russula amoenolens* Romagnesi

Identification marks   At first the cap is viscid, cushion-shaped, and evenly dark grayish brown; as it matures, it becomes plane to slightly depressed in the center, remains viscid, and fades and becomes tuberculate-striate toward the margin. The gills are distinctly acrid. The odor is waxy to woodsy.

Edibility   Not recommended.

When and where   Solitary to gregarious under pines, oaks, and in mixed woods; summer and fall; widely distributed in eastern North America.

Microscopic features   Spores 6–7.5 × 5–6.5 μm, ornamentation in the form of isolated warts to 1 μm high. Cuticle of cap a thick ixocutis with scattered clusters of hyphal ends on it.

Observations   The specific epithet means fragrant or sweet smelling. *R. subfoetens*, *R. pectinatoides*, and *R. amoenolens* are all members of subsection Foetentinae and are closely related.

113 *Russula amoenolens*   About one-half natural size

Russula pectinatoides Pk.   **114**
(Pectenlike Russula)

Identification marks   At first the caps are viscid, radially striate, and light grayish yellowish brown; in age they remain striate but become dry and pale straw color to yellowish tan or grayish yellowish brown. The gills are yellowish white when young and become pale yellow as the spores mature. In contrast to the somewhat acrid taste of the gills, the cuticle of the cap and its context are mild. The odor is somewhat unpleasant. The spore deposit is pale orange yellow.

Edibility   According to Peck it is "edible but not highly flavored."

When and where   Solitary to gregarious in deciduous and mixed forests, under pines with an understory of hardwoods, and in shaded lawns and parks; summer and fall; described from New York and widely distributed east of the Great Plains, common in the South.

Microscopic features   Spores 5.5–8.5 × 4.5–6 μm (excluding ornamentation); ornamentation to 1 μm high as blunt, predominantly isolated warts. Cuticle of cap a thick ixocutis that gives rise to a discontinuous ixotrichoderm.

Observations   The specific epithet means resembling *R. pectinata*; *pectinatus* means having narrow close divisions like a comb, in this case a reference to the striate margin of the cap.

114 *Russula pectinatoides*   About two-thirds natural size

137

Hygrophoraceae

Among the gilled mushrooms with white spores in deposits, fleshy stalks, and attached gills, the Hygrophoraceae stand out because of the "waxy" texture and appearance of the gills. Detecting this character may be a difficult problem for beginners, but with practice it becomes easier. The gills look clean, waxlike, and are seldom crowded. The same texture, it must be admitted, also occurs in *Laccaria* in the Tricholomataceae. Technical characters of the family include smooth inamyloid spores (in most species) that lack an apical pore, and long, narrow basidia. Genera within the family are defined primarily on the arrangement of hyphae in the gills. Most members of this family occur on humus where they apparently form mycorrhizae with various plants. The name *Hygrophorus* means moisture bearing or damp, *Hygrocybe* means damp or wet head or cap.

Key to Species

1. Stalk 8–20 mm thick near apex, white or nearly so 2
1. Stalk 2–6 mm thick near apex, brightly colored 3
    2. Cap olive brown when young becoming yellowish brown and finally yellow . . . . 115. *Hygrophorus hypothejus*
    2. Cap pinkish buff to pinkish brown 116. *Hygrophorus roseibrunneus*
3. Stalk thinly slimy; entire fruiting body yellow, cap fading in age . . .
    . . . . 117. *Hygrocybe nitida*
3. Stalk moist to dry; entire fruiting body orange when young, cap and stalk fading in age but gills long remaining bright orange . . . .
    . . . . 118. *Hygrocybe marginata*

---

115   *Hygrophorus hypothejus* (Fr. : Fr.) Fr.

Identification marks   In button stages a veil of slime envelops the entire fruiting body. Young caps are dark olive brown to yellowish brown, as they age they become yellow to orange or even red. A thin zone of fibrils on the stalk is usually all that remains of the fibrillose inner veil. The gills are white at first and pale yellow to orange in age.

Edibility   Edible but not popular because of the dirt that adheres to the specimens.

When and where   Gregarious to scattered under conifers, particularly pines, after the onset of cool weather in the fall and in the winter; widely distributed in North America.

Microscopic features   Spores 7.5–9 × 4–5.5 μm. Cuticle of cap an ixolattice to an ixocutis.

Observations   This is a common winter mushroom in parts of the South with mild winters. The fruiting bodies vary from robust to slender. The specific epithet is generally translated as meaning "under brimstone." Its relevance to this species is obscure.

115 *Hygrophorus hypothejus*   About natural size

Hygrophorus roseibrunneus Murr.   **116**

Identification marks The gills and stalk are white and the cap is pinkish cinnamon to pale pinkish buff or pinkish brown. In wet weather both the cap and stalk are sticky to slimy. The stalk is pruinose or dotted with clusters of fibrils in addition. Neither the odor nor the taste are distinctive.

Edibility Edible.

When and where Scattered to gregarious under hardwoods and/or conifers; late fall into winter. It was described from California and is not uncommon in the South.

Microscopic features Spores 7.5–9 × 4–5 µm. Cuticle of cap a thin ixocutis to ixolattice.

Observations *Roseibrunneus* means rose brown.

116 *Hygrophorus roseibrunneus*   About two-thirds natural size

117 *Hygrocybe nitida* (Berk. & Curt.) Murr.

Identification marks   The cap and stipe are slimy when fresh, all parts of the fruiting body are yellow (never orange), and the cap becomes whitish when faded. The gills are narrow, yellow, and decurrent. The stalks are 2−5 mm thick and do not discolor at the base from handling.

Edibility   Edible but not popular because of the small size of the fruiting bodies and their slimy surfaces.

When and where   Gregarious to clustered on soil in moist woods and in or along the edges of bogs under broad-leaved trees; summer and early fall. It was described from South Carolina and is fairly common and widely distributed east of the Great Plains.

Microscopic features   Spores 6.5−8 (9) × 4−5 (6) µm. Cuticle of cap an ixocutis.

Observations   *Nitidus* means shining or polished. The shining appearance of these fruiting bodies is caused by the thin layer of slime on the cap and stalk.

117   *Hygrocybe nitida*   Natural size

118 *Hygrocybe marginata* (Pk.) Murr.

Identification marks   The persistently bright orange gills are the most reliable way of identifying this species. At first the cap is moist and orange as is the stalk but they fade as they age. The odor and taste are mild.

Edibility   We have no data on it. It is seldom found in sufficient quantity to be eaten but does attract attention because of the bright orange gills.

When and where   Widely distributed in eastern North America in mixed woods and occasional in the West; summer and fall; regularly encountered but seldom abundant.

Microscopic features   Spores 7−10 × 4−6 µm. Cuticle of cap a cutis.

Observations   The specific epithet means with a (colored) margin.

118 *Hygrocybe marginata*

Natural size

Amanitaceae

Either the gills do not touch the stalk (are free) or they are scarcely attached to it. The spores are white in deposits. A universal (outer) veil is regularly present, and a partial veil (one extending from the stalk to the cap margin) may or may not be present. The spores are smooth and lack an apical pore.

The two genera in this family most likely to be encountered in the South are *Amanita* and *Limacella*.

Key to Genera

1. Outer veil slimy . . . . (p. 141) *Limacella*
1. Outer veil powdery, membranous, cottony, or fibrillose but never slimy . . . . (p. 143) *Amanita*

Limacella

The caps are glabrous, viscid when young, and lack superficial veil remnants such as are present in many species of *Amanita*. The stalks are either dry or viscid when young.

This is a genus of about a dozen species that fruit on the ground or on very decayed wood. The genus name seems to be derived from a word for mud or slime.

Key to Species

1. Fruiting body white all over, slime colorless . . . . . . 119. *Limacella illinita*
1. Cap golden brown to dull yellow beneath the pale yellow to ochraceous slime . . . . 120. *L. kauffmanii*

119 *Limacella illinita* (Fr. : Fr.) Earle

Identification marks Among the numerous white mushrooms, this one can be recognized by the combination of slimy cap and stalk (actually the remains of a gelatinous universal veil); lack of both a volva and a conspicuous annulus; white spore deposit; and free, white gills. The slime is clear and copious; in wet weather it may drip off the cap. Both the odor and taste are mild.

Edibility Reported to be edible, but the soft flesh and slimy veil limit its appeal.

When and where Solitary to gregarious in shaded lawns and in oak woods; summer and fall; widely distributed in North America and common and often abundant in the South.

Microscopic features Spores 5–6.5 × 4–4.5 μm, inamyloid. Cuticle of cap an ixotrichoderm to ixolattice of narrow hyphae in a thick matrix of slime.

Observations *Illinitus* means smeared. The variety shown is var. *illinita*; in var. *rubescens* reddish stains develop on the stalk; and in var. *argillacea* the center of the cap and the stalk are tinged with dull brown.

119 *Limacella illinita*

Natural size

120 *Limacella kauffmanii* H. V. Smith

Identification marks Beneath the slime the cap is golden brown to dull yellow. Both the cap and stalk are covered with pale yellow to ochraceous slime. The taste varies from mild to somewhat meallike (farinaceous).

Edibility Not reported.

When and where Solitary to gregarious or loosely clustered on the ground under deciduous trees; summer and early fall; described from Chain Bridge, Virginia, and widely distributed in the South but often overlooked.

120 *Limacella kauffmanii*   About one-third natural size

Microscopic features   Spores 3–5 × 3–3.5 µm. Cuticle of cap an ixotrichodermium collapsing to form an ixolattice.

Observations   The late C. H. Kauffman (1869–1931) was a professor of
botany at the University of Michigan and was the first to collect this species.
It is close to *L. glishera*, which has bright reddish brown slime.

Amanita

Amanita is one of the larger genera of the Agaricales. In the South
both excellent edible and deadly poisonous species of *Amanita* occur. No one should eat any member of this genus without first becoming thoroughly familiar with the species in the field and through
technical characters. It is important to be critical in interpreting
characters—your life may depend on it. All but a few species "look
like an *Amanita*" once one knows the general features of the genus.
Because of the serious consequences that could result from mis-
identifications and the fact that much remains to be learned about
this genus in North America, we categorically do not recommend
eating any species of *Amanita*. Missing out on a few meals of wild
mushrooms is a small price to pay to avoid a potentially serious
case of poisoning.

The universal or outer veil and what becomes of it provide many
of the important characters used to distinguish species in this
genus. This veil envelops the button mushrooms, as the buttons
expand, it is ruptured and its remnants may take a variety of forms
and positions depending on the species. In some species this veil
leaves a cup of tissue, called the volva, at the base of the stalk,
e.g., *A. bisporigera*, in others, e.g., *A. muscaria*, it leaves bands of
soft tissue along the stalk; or it may not leave any trace at all on the
stalk as in many collections of *A. rubescens*. On the cap it may not
leave any trace at all or it may persist as patches of tissue, distinct
warts or spines, or even a soft powder.

A partial veil is present in addition to the universal veil in some
species. It extends from the edge of the cap to the stalk in young

specimens. When it ruptures it often leaves a distinct skirtlike or membranous annulus on the stalk. The annulus may be persistent or evanescent in which case all traces of it soon vanish. In a large number of species the remnants of the partial veil adhere in patches along the cap margin, such a margin is said to be appendiculate. Those species in which the cap margin is appendiculate and the spores are amyloid are placed in the subgenus *Lepidella*. In North America, this subgenus reaches its greatest diversity in the South and is represented in this book by *A. cokeri*, *A. abrupta*, *A. hesleri*, *A. onusta*, *A. praegraveolens*, *A. polypyramis*, and *A. thiersii*.

Other characters the student of this genus will have to use include whether the spores are amyloid or inamyloid, whether or not clamp connections are present, and the shape and arrangement of the cells and hyphae in the universal veil.

The name of the genus is derived from a word that was in use at least as early as classical times for a kind of fungus.

Key to Species

13. Cap glabrous, margin not appendiculate at first . . . . 131. *A. bisporigera*
13. Cap with patches of universal veil remnants at first, margin appendiculate in young specimens . . . . 132. *A. mutabilis*
    14. Cap bright red to orange fading to yellow, decorated with distinct warts . . . . 133. *A. muscaria*
    14. Cap white or not colored as above 15
15. Center of young cap decorated with chunky warts, remainder covered with a powdery orange to pinkish salmon veil; chunky warts lost by maturity leaving a characteristic "bald spot" in the center of the cap . . . . 134. *A. roseitincta*
15. Not as above 16
    16. Fruiting body white to ivory overall 17
    16. Either fruiting body pale pinkish tan or scales and/or warts on the cap distinctly colored 21
17. Base of stalk a somewhat pointed bulb with coarse warts and recurved scales that are not easily brushed off . . . . 135. *A. cokeri*
17. Not as above 18
    18. Stalk slender, only slightly bulbous at the base, covered with lacy, often sticky veil remnants . . . . 136. *A. thiersii*
    18. Not as above 19
19. Veil remnants on cap in the form of membranous patches . . . .
. . . . see 132. *A. mutabilis*
19. Veil remnants in the form of warts, powder, or scales 20
    20. Base of stalk abruptly enlarged to form a subglobose bulb . . .
. . . . 137. *A. abrupta*
    20. Base of stalk gradually enlarged toward the base . . . .
. . . . 138. *A. polypyramis*
21. Cap and stalk light pinkish tan . . . . 139. *A. praegraveolens*
21. Not as above 22
    22. Veil remnants on cap in the form of powdery gray warts and granules . . . . 140. *A. onusta*
    22. Veil remnants on cap in the form of more or less concentric rings of flat brown scales . . . . 141. *A. hesleri*

Amanita farinosa Schw.   **121**

Identification marks   The crustlike patches on the cap and striate margin together with the stout stature of some specimens may remind one of certain species of *Russula*. The free gills and remnants of the veil on the cap and base of the stalk indicate that it is an *Amanita*. There is no annulus. The spores are not amyloid; those of all species of *Russula* are. The cap is tan to light brown under the gray veil remnants.

Edibilty   We have no information on it; not recommended.

When and where   Solitary to scattered in thin woods; summer and fall; widely distributed in eastern North America and most abundant in the Southeast. Schweinitz described it from North Carolina.

Microscopic features   Spores 7–8 × 5.5–7 μm, inamyloid. Clamp connections not observed. Veil remnants on cap predominantly of globose to subglobose cells.

Observations   *Farinosus* means mealy or covered with meal. This species can be quite "un-*Amanita*-like," but a check of the spores (scraped up

121 *Amanita farinosa*   Natural size

from a deposit and tested with iodine or Melzer's reagent on a slide or plate) will prove it is not a *Russula* in spite of its resemblance to *R. pectinatoides* and *R. balloui*.

122   *Amanita ceciliae* (Berk. & Broome) Bas
(Gray Amanitopsis)

Identification marks   Gray warts or patches of veil tissue persist at maturity on the gray brown to gray cap. The margin of the cap is slightly striate. No annulus is present on the stalk, and the base of the stalk is not noticeably enlarged or bulbous. The pale gray volva fits tightly around the stalk, leaving no space between them. The rim of the volva is often darker gray than the remainder and forms a ringlike zone near the base of the stalk.

122 *Amanita ceciliae*   One-half natural size

Edibility   Reported to be edible, but not recommended.

When and where   Common and widely distributed, often scattered in grassy places near or under pines as well as in mixed woods; from spring until the end of the growing season.

Microscopic features   Spores (10) 12–15 μm in diameter, basically globose, inamyloid. Veil remnants on cap of hyphae with both narrow and greatly inflated cells. Clamp connections not observed.

Observations   Like *A. vaginata* this is a "collective" species that is probably an aggregation of distinct taxa which have yet to be sorted out. This species is also known as *A. strangulata* and *A. inaurata*; however, the name used here has been shown to be correct for this species. The specific epithet honors Cecilia E. Berkeley.

123   *Amanita spreta*   Two-thirds natural size

Amanita spreta (Pk.) Sacc.   **123**

Identification marks   In our opinion, this is a collective species in need of detailed study. The basic features are the off-white to pale or dark gray to grayish brown glabrous cap; the thin annulus that often collapses by maturity; and saccate membranous volva. The base of the stalk is not at all or only slightly enlarged. At maturity the margin of the cap is striate.

Edibility   Some variants are nontoxic, but the edibility of others has not been reported; not recommended.

When and where   Gregarious or sometimes in fairy rings under hardwoods and in mixed woods, often on sandy soil. The best fruitings occur during hot humid weather. Peck described it from eastern New York and it is widely distributed east of the Great Plains.

Microscopic features   Spores 10.5–13.5 × 6–8 μm, inamyloid. Clamp connections present.

Observations   The specific epithet means hated or despised. The specimens shown here are a pale variant in which the annulus is evanescent and the striations on the cap distinct.

124 *Amanita vaginata* (Bull. : Fr.) Quél. (Grisette)

Identification marks These dull gray caps are striate from the margin almost to the disc, umbonate in age, and typically lack veil remnants. There is no annulus. The volva is white, sacklike, and does not adhere to the stalk—which is neither strongly swollen nor bulbous at the base.

Edibility Reported to be edible, but everyone who has not studied the genus carefully should avoid it.

When and where Scattered to gregarious, sometimes in troops, under pines and/or hardwoods; summer and fall; widely distributed and common in many of the forested areas of North America.

Microscopic features Spores 10.5–13.5 µm in diameter, basically globose, inamyloid. Clamp connections not observed.

Observations Mycophagists should be very careful with their identifications if they decide to eat this species as there are several gray species of *Amanita* in the South (see also *A. ceciliae*). The specific epithet means sheathed. There are several taxa with the same basic field characters that vary in cap color from white to gray, tan, or orange. One specialist on *A. vaginata* and its relatives has said "the available information is a mess" so if you are confused, know you are in good company.

124 *Amanita vaginata*   About two-thirds natural size

125 *Amanita umbonata* Pomerleau

Identification marks In buttons the cap is brilliant red, when fully expanded it is red to orange red over the disc and yellow at the margin. Both a persistent, membranous annulus and a tough, white, sacklike volva are

125 *Amanita umbonata*   About one-third natural size

present. The stalk and gills are pale yellow. Usually no remains of the veils persist on the cap.

Edibility Edible and good but recommended only for those willing to make a careful study of the genus.

When and where Scattered to gregarious in mixed woods, often under pine or oak; summer and fall; widely distributed and common in the South. It is also common in the St. Lawrence Valley of Canada.

Microscopic features Spores 8–11 (12) × 6–8 μm, inamyloid. Clamp connections present.

Observations The specific epithet is a reference to the prominent umbo (raised, blunt bump) in the center of mature caps. Until recently this species has been called *Amanita caesarea* (Caesar's mushroom) in North America.

Amanita gemmata (Fr.) Bertillion in Dechambre   **126**

Identification marks These rain-washed specimens have lost all traces of the universal veil from the caps and serve as a reminder that not all specimens of a species show all the characters of the species at the time they are found. The pale but distinctly yellow, viscid cap that is striate near the margin; thin partial veil that may or may not form an annulus; and bulbous base of the stalk with a membranous rim on the volva are important characters.

Edibility Since there are contradictory reports on the chemical constituents of this species, we do not recommend it for table use.

When and where It fruits in summer and fall in many of the forested regions of North America and is locally and sporadically abundant on sandy soil under mixed hardwoods in the South.

Microscopic features Spores 9–11 × 6–7 μm, inamyloid. Clamp connections present.

126 *Amanita gemmata*   Natural size

Observations   The specific epithet means provided with buds, probably a reference to the warts usually left by the universal veil on the cap. The group of variants around this species is a complex one here in North America.

127   *Amanita flavorubens* (Berk. & Mont.) Sacc.

Identification marks   The cap is yellow to orange yellow or yellowish brown and decorated with firm pale yellow warts. The annulus is skirtlike, persistent, and often has a yellow edge. At most only a few inconspicuous

127 *Amanita flavorubens*   About one-half natural size

patches of tissue persist as the volva at the base of the stalk. In age dull red stains develop on and in the base of the stalk, often around worm holes.

Edibility   Not recommended; although it is generally reported to be poisonous or suspected, we know of no documented poisonings by it.

When and where   It is widely distributed in the forested regions of eastern North America. Although it was described from Ohio, it is most abundant in the South. Summer and fall are the best times to look for it.

Microscopic features   Spores 7.5–10.5 × 5.5–7.5 μm, amyloid. Clamp connections not observed. Veil remnants on cap composed of hyphae with cells of many shapes, some with pale tan content.

Observations   This species has long been called *A. flavorubescens* in North America. The name *A. flavorubens* was proposed forty-six years earlier for what we believe is the same species and thus has priority and should be used.

128   *Amanita rubescens*   About one-half natural size

Amanita rubescens (Pers. : Fr.) Pers.   **128**
(The Blusher)

Identification marks   Small, firm, whitish to yellowish or rusty brown warts are scattered over the cap. The color of the cap varies greatly from collection to collection and is usually some shade of pale pinkish tan, olive brown, or reddish brown. On the stalk there is a persistent annulus but no volva. All parts of the fruiting body stain purplish red (vinaceous red) in age or when injured, not just the base of the stalk as in *A. flavorubens*.

Edibility   Edible and good but recommended only for those who have studied the genus thoroughly.

When and where   Solitary to gregarious under hardwoods and in mixed woods, especially under oaks; summer and fall; common east of the Great Plains and often abundant in the South.

Microscopic features   Spores 7–9 × 4.5–6 μm, amyloid. Clamp connections not observed. Veil remnants on cap a mixture of greatly inflated cells and filamentose hyphae.

Observations   *Rubescens* means becoming red, an appropriate name for this species. As is the case with many common species, *A. rubescens* is really a collective species.

129 *Amanita volvata* (Pk.) Lloyd

Identification marks Inconspicuous soft floccose patches of veil remnants are randomly distributed over the cap. It is faintly striate near or at the margin and whitish to cream colored with the disc varying to pale brown. No annulus is present. At first the fibrils on the stalk are white but in age they become rusty brown. The outstanding feature, however, is the tough, large, white, rounded volva that has a free lobed margin.

Edibility Not reported and not recommended.

When and where Scattered in woods; summer. Peck gathered it in New York, the specimens in our illustration are from Mississippi, and it occurs at least as far west as eastern Texas.

Microscopic features Spores 7.5–10.5 × 4.5–6 µm, amyloid. Clamp connections not observed. Remnants of veil on cap of hyphae whose cells are slightly inflated but not globose.

Observations Peck noted that the gills of dried specimens were pale cinnamon rather than whitish as in many members of this genus. The left-hand cap in our illustration is partly covered with spores from another specimen and shows one way one can often determine spore color. *Volvatus* means provided with a volva.

129 *Amanita volvata*   About three-fourths natural size

130 *Amanita citrina* (Schaeff.) Pers.

Identification marks The large rounded bulb is often cleft like a loaf of bread that was slashed then allowed to rise a second time. There is a persistent membranous annulus on the stalk. The cap is thinly slimy and varies

130 *Amanita citrina*   Two-thirds natural size

from pale whitish yellow to dull yellow or greenish yellow. White, pale yellow, or more often grayish to lavender gray feltlike patches of tissue are scattered over the cap. The odor is reminiscent of raw potatoes.

Edibility Not recommended because of the danger of confusing it with *A. phalloides*; furthermore, the flavor is said to be unpleasant.

When and where Widely distributed in woods of various types east of the Great Plains; from spring until the onset of cold weather; sporadically abundant.

Microscopic features Spores 7.5–11 × 7.5–10.5 μm, amyloid. Clamp connections not observed. Veil remnants on cap composed of large, predominantly ellipsoid cells.

Observations The specific epithet means citron-colored or greenish yellow. *A. phalloides*, one of the most poisonous species in the genus, can be similar in color but differs in having a distinct cuplike volva at the base of the stalk, and the cap is usually darker, more of a dull olive.

131

Amanita bisporigera Atk.
(Smaller Death Angel)

Identification marks All parts of these slender fruiting bodies are pure white at first; the partial veil forms a persistent, skirtlike annulus; and the base of the stalk is set in a membranous volva which has a free margin. In age the cap may be tinged with pale tan or creamy yellow. No veil remnants persist on the cap.

Edibility Dangerously, and sometimes deadly, poisonous; amatoxins (see p. 11) are the principal toxins.

When and where Scattered in deciduous and mixed woods and wooded lawns especially near oaks; late May through fall. Described from Ithaca, New York, this species is widely distributed east of the Great Plains and common.

Microscopic features Spores 9–11 × 7.5–9 μm, amyloid. Clamp connections not observed. Basidia 2-spored.

153

131   *Amanita bisporigera*   One-half natural size

Observations   The specific epithet refers to the 2-spored basidia; in the other white species of *Amanita* that are superficially similar, most basidia are 4-spored. *A. virosa*, the death angel, has larger, fleshier fruiting bodies, and spores that are 8–11 × 8–11 μm. It is also abundant in the South; however, it starts to fruit somewhat later than the smaller death angel. All mycophagists should learn these species in order to avoid eating them. The cuplike volva of these species is often termed a death cup.

132   *Amanita mutabilis* Beards.

Identification marks   A typical specimen of this species has a fleshy, abruptly enlarged stalk base with the free edge of the volva standing away from the stalk; a smooth, whitish to pale pink cap; evanescent annulus; and

132   *Amanita mutabilis*   About natural size

154

thick gills. Bruised areas may become bright pink to dusky rose. Some collections have an aniselike odor. The underside of the partial veil is roughened as if it had been dusted with coarse meal unlike *A. bisporigera* in which it is more or less smooth.

Edibility Unknown and not recommended; it is easily mistaken for *A. virosa.*

When and where It fruits from summer until the end of the growing season in a solitary to scattered manner in mixed woods and under pine. Davis Island, North Carolina, is the type locality and the limits of its known distribution include New Jersey, Florida, and Texas.

Microscopic features Spores 11–13.5 (15) × 6.7–7.5 µm, amyloid. Clamp connections absent. Remnants of veil on cap of narrow hyphae and scattered inflated, mostly ellipsoid, cells.

Observations *Mutabilis* means changeable. An intriguing character of this species is that young basidia contain numerous reddish (dextrinoid) granules when examined in Melzer's reagent.

133   *Amanita muscaria* var. *flavivolvata*   About one-third natural size

Amanita muscaria var. *flavivolvata* (Sing.) Jenk.   133
(Fly Agaric)

Identification marks Young caps are bright red to orange red; old ones fade to dull orange or paler. The cap is studded with numerous, firm, blunt, yellow to pale orange yellow warts. Both veils, particularly the underside of the partial veil, are pale yellow to orange yellow. In age the annulus often collapses and disappears. The base of the stalk consists of a rounded to egg-shaped bulb with zones of firm recurved scales that may extend up the stalk for a short distance. The veil remnants may fade to white in age.

Edibility Poisonous; like other varieties of this species, it presumably contains ibotenic acid (see p. 11).

When and where Often in arcs or fairy rings, under conifers, hardwoods, and in mixed woods; late fall. This is the common variety of *A. muscaria* in the coastal plain; it also occurs in California as well as Mexico and Guatemala.

Microscopic features Spores 9–11 × 7–8 μm, inamyloid. Clamp connections present. Veil remnants on cap composed of subcylindric to elliptic cells and binding hyphae.

Observations The yellow volval remnants and evanescent annulus serve to distinguish this variety in the field from var. *muscaria* in which the veils are white and the annulus persistent. One or more varieties of this species can be found in almost every forested part of North America. *Musca* means a fly; *flavivolvata* means yellow volva.

---

134 *Amanita roseitincta* (Murr.) Murr.

Identification marks Fresh young buttons have a patch of chunky warts in the center of the cap that soon fall away and leave a bald spot. The rest of the cap is covered by a thin, powdery, dull brownish orange to pinkish tan coating which may be almost invisible in age. The cap itself is creamy white to pale buff. The lower surface of the thin evanescent annulus is colored like the coating on the cap. Only a few chunks of tissue and/or some colored powder represent the volva at the base of the stalk.

Edibility Not determined and not recommended.

When and where It fruits during hot damp weather in midsummer and early fall under pines, hardwoods, and in mixed woods. Murrill described it from a collection made by Esther S. Earle (Mrs. F. S.) in Biloxi, Mississippi. We have found it several times in Mississippi and expect it is not uncommon in the coastal plain.

Microscopic features Spores 9–10.5 (12) × 6–7.5 μm, inamyloid. Clamp connections not observed. Powdery coating on cap composed of scattered to clustered, globose, subglobose, or ellipsoid cells.

134a *Amanita roseitincta*, button

Natural size

134b *Amanita roseitincta*, mature specimens   One-third natural size

Observations Mature specimens might be mistaken for peculiar specimens of *Lepiota naucinoides* unless one notes the powdery veil remnants
on the cap and base of the stalk. As expected, the specific epithet means
rose-tinted.

Amanita cokeri (Gilb. & Kühner) Gilb.   **135**

Identification marks At the time the veil breaks, the margin of the cap is
decorated with numerous fragments of the veil. The cap is white, rather viscid, and beset with soft conical warts. Furthermore, the annulus is large and
attached near the apex of the stalk, and the bulb at the base of the stalk is
ventricose. These features distinguish it from white specimens of *A. muscaria* in the field. The amyloid spores are the most reliable distinction,
however.

135 *Amanita cokeri*   One-fourth natural size

Edibility   We have no information on it; not recommended.

When and where   Scattered to gregarious in pine duff and in mixed woods from summer through fall. It was described from North Carolina and is expected to occur throughout the South.

Microscopic features   Spores 10.5–13.5 × 7–9 μm, amyloid. Clamp connections present. Veil remnants on cap a mixture of binding hyphae and globose to broadly ellipsoid cells.

Observations   W. C. Coker, who studied the amanitas of North Carolina, is commemorated by the specific epithet.

---

136   *Amanita thiersii* Bas

Identification marks   The long, slender, shaggy stalk appears to be covered with torn lace which sticks to one's fingers when touched. The base of the stalk is slightly swollen and lacks a membranous volva. Wooly to fibrillose soft patches of tissue, scales, and warts nearly cover the cap and often hang from the margin of young specimens. The annulus is frequently collapsed by maturity.

Edibility   We have no information on it; not recommended.

When and where   In arcs or fairy rings in lawns and open areas and in open woods; known so far only from Texas and Mississippi but to be expected in the coastal plain generally; August and September.

Microscopic features   Spores (7.5) 9–10.5 × 7.5–9 μm, amyloid. Clamp connections not observed. Veil remnants on cap composed of hyphae with elongate to sausage-shaped inflated cells mixed with some binding hyphae.

Observations   This species was named for H. D. Thiers of San Francisco State University who collected it in College Station, Texas. Fairy rings of it that intersect fruitings of edible species are not uncommon and emphasize the need to check the characters of each specimen gathered for food.

---

137   *Amanita abrupta* Pk.

Identification marks   The entire fruiting body is pure white or the center of the cap may be pale straw yellow. Small, pointed, superficial scales, often in concentric rings, cover the cap when fresh but may be washed away by rain. A large, membranous, persistent annulus is present. Most striking of all is the abruptly enlarged basal bulb for which the species is named. At most only a low ridge of tissue above the bulb or on it is left as a volva.

Edibility   Reported to be edible, but we *do not* recommend experimenting with any white *Amanita*.

When and where   Scattered to gregarious under hardwoods or in mixed woods, at times abundant; summer and fall. It was described from specimens gathered in Alabama by Underwood and occurs primarily in the eastern and southern United States but it has been reported as far west as Iowa.

Microscopic features   Spores 7.5–9 × 6–7.5 μm, amyloid. Clamp connections present. Veil remnants on cap a mixture of clavate, ovate, and elongate cells with some binding hyphae.

Observations   In young specimens the margin of the cap is appendiculate, a characteristic of a large group of species in the genus.

136 *Amanita thiersii* One-fourth natural size

137 *Amanita abrupta* One-half natural size

Amanita polypyramis (Berk. & Curt.) Sacc. **138**

Identification marks The fruiting bodies are large, stout, and white. Young caps are beset with small pointed warts in the center that vary to flat scales or thin patches of tissue near the margin that is hung with bits of tissue. There is a powdery coating over much of the stalk above the large bulb, which bears small pyramidal warts on its upper portion. Chunky patches of tissue are present on the lower surface of the intact partial veil which usually falls away without forming an annulus.

Edibility Not recommended. Old specimens have a strong odor variously described as "old ham," "peculiar and penetrating," and "chloride of lime"— hardly appetizing.

When and where Solitary to scattered in dry pine-oak woods; October into December; mid-Atlantic states south into Florida and west into Texas; Curtis found it in South Carolina.

138 *Amanita polypyramis*

One-third natural size

Microscopic features Spores 10–12 (13) × 6–7.5 μm, amyloid. Clamp connections not observed. Veil remnants on cap composed of a mixture of binding hyphae and inflated, globose to predominantly elongate cells.

Observations The appearance of these large white mushrooms is a sign that the fall mushroom season is underway. The specific epithet means many pyramids.

139 *Amanita praegraveolens* (Murr.) Sing.

Identification marks Light pinkish tan is the predominant color of the caps in our experience although white fruiting bodies have been reported. At first the surface of the cap is blocked out into low areolae but at maturity it is distinctly scaly. Remnants of the veil adhere to the cap margin and may also form an irregular annulus. There is no membranous volva and the base of the stalk is only slightly if at all bulbous. The odor is strong and unpleasant.

Edibility Not reported and not recommended.

When and where Often in fairy rings in lawns and mowed areas as well as scattered under hedges; summer and fall. The type collection was made in Gainesville, Florida; we suspect the species occurs generally around the Gulf of Mexico.

Microscopic features Spores 7–9 × 6.5–7.5 μm, amyloid. Clamp connections present. Volval remnants on cap of hyphae with elongate, inflated cells.

Observations The specific epithet means very strong smelling.

139 *Amanita pragraveolens*

One-fourth natural size

Amanita onusta (Howe) Sacc.   **140**

Identification marks   The crowded, more or less conical warts on the cap and similar warts and scales on the stalk are important. The base of the stalk is long, spindle-shaped, and often deeply inserted in the ground. The fruiting bodies are small to medium in size. In age, when the warts have worn off, the cap appears powdery.

Edibility   Undetermined but not recommended.

When and where   To be expected in sandy soil under birch along rivers and streams; summer and fall; reported from northeastern North America south into the mountains of western North Carolina.

Microscopic features   Spores 9–11 × 6–8 µm, amyloid. Clamp connections present. Veil remnants on cap composed of slender binding hyphae and upright rows of subglobose to broadly ellipsoid cells with light gray brown contents.

Observations   *Onustus* means overloaded or full, perhaps a reference to the numerous warts on the cap.

140   *Amanita onusta*   About three-fourths natural size

Amanita hesleri Bas   **141**

Identification marks   The universal veil leaves soft, fibrillose, gray brown to dark brown, appressed scales on the cap; at times low warts may be present over the disc. Brown to gray brown bits of tissue represent the volva around the somewhat swollen base of the stalk. At most only a few wisps of tissue persist as an inconspicuous annulus.

Edibility   We have no information on it; not recommended.

When and where   Scattered under pines and in mixed woods; summer and fall; uncommon to rare. It was described from specimens collected by W. C. Coker in Chapel Hill, North Carolina; Hesler gathered it in Tennessee; our specimens were found in Mississippi.

141 *Amanita hesleri*   Two-thirds natural size

Microscopic features Spores 10.5–12 × 5.5–7 μm, amyloid. Clamp connections not observed. Veil remnants on cap of clavate and ellipsoid cells.

Observations It is a pleasure to illustrate this species named for L. R. Hesler, one of the South's great mycologists. Superficially it resembles *Armillaria caligata* in having soft brown scales on the cap; however, the *Armillaria* has a stalk that tapers toward the base and is covered with bands of brown fibrils below a distinct annulus; furthermore, the gills are attached to the stalk.

Lepiotaceae

There is a superficial similarity between members of this family and some species of *Amanita*, particularly members of subgenus *Lepidella*. In both groups the gills are free and the spores, except in *Chlorophyllum*, are white or nearly so in deposit. An annulus is typically present, at least for a limited time, in the Lepiotaceae. The important difference that can be used to distinguish the two families without recourse to a microscope is that a volva and the other types of evidence of a universal veil present in the Amanitaceae are absent from the Lepiotaceae. When in doubt as to which family is involved, use a microscope and check the arrangement of the hyphae in the gills. In the Amanitaceae the hyphae diverge from the middle of the gill down and out toward the two sides; in the Lepiotaceae they are interwoven to parallel throughout. Furthermore, none of the species in the Lepiotaceae that have large fruiting bodies have amyloid spores as do many species of *Amanita*.

There are many members of this family in the South, only a few of which are covered here. We recognize three genera: *Chlorophyllum*, *Lepiota*, and *Leucocoprinus*. The names mean green leaf (or gill, in this case); having scales or scaly ear; and white or pale *Coprinus* respectively.

Key to Species

1. Spores dull green in deposits, gills dull green in age; often fruiting in arcs or circles in lawns and open areas ....
    .... 142. *Chlorophyllum molybdites*
1. Not with the above combination of characters 2

2. Bruised areas on the cap and stalk becoming dull red, purple, or brown with or without a preliminary change to yellow 3
2. Injured areas unchanging or not changing as above 5
3. Disc and scales on cap dingy yellow to yellowish brown, stalk equal in diameter above a slightly bulbous base . . . . 147. *Lepiota humei*
3. Not as above 4
  4. Disc and scales on cap very dark brown to nearly black . . . . .
     . . . . 143. *Lepiota sanguiflua*
  4. Disc and scales on cap cinnamon brown or pinkish brown . . .
     . . . . 144. *Lepiota besseyi*
5. Fruiting body often stout and squat; cap white and smooth at all stages, margin not striate . . . . 145. *Lepiota naucinoides*
5. Not as above 6
  6. Annulus with a thick ragged edge; cap medium to large (over 6 cm broad) and fleshy 7
  6. Annulus thin, membranous, sometimes evanescent; cap small to medium (less than 5 cm broad) and thin fleshed 8
7. Scales and scurf on cap and stalk light to medium grayish cinnamon; fruiting in or at the edge of woods . . . . 146. *Lepiota procera*
7. Scales on cap dull dingy yellow to tan; stalk smooth; often fruiting in lawns, pastures, and open areas . . . . 147. *Lepiota humei*
  8. Cap yellow 9
  8. Cap white, tan, or white with dull purple disc 10
9. Cap almost translucent; fruiting bodies extremely fragile (breaking readily when touched or picked); fruiting in woods . . . .
   . . . . 148. *Leucocoprinus fragilissimus*
9. Cap somewhat fleshier; fruiting bodies firmer; fruiting in lawns, gardens, mulches, etc. . . . . 149. *Leucocoprinus luteus*
10. Cap pure white . . . . 150. *Leucocoprinus breviramus*
10. Cap at least in part colored 11
11. Cap light to medium tan overall . . . . . . 151. *Leucocoprinus longistriatus*
11. Cap with a dull purple to purplish brown disc and scales . . . .
   . . . . 152. *Leucocoprinus lilacinogranulosus*

Chlorophyllum molybdites (Meyer) Massee   142
(Green-gilled Lepiota)

Identification marks The outstanding features of this species are the gills that become dingy olive green as the spores mature and the spores that are dull olive green to grayish green in deposits. In buttons the caps are covered with a pale pinkish tan, smooth "skin" (cuticle) that remains intact over the disc and breaks up into scales and patches toward the margin. The annulus has a thick, ragged edge, and is moveable on old or partially dried specimens.

Edibility Poisonous for most people; it may cause nausea, vomiting, cramps, and diarrhea. According to some reports the toxins are either removed or destroyed by boiling, which may account for some of the variability found in reports on its edibility.

When and where Gregarious, often in arcs or fairy rings in grassy areas, particularly lawns. It fruits from early summer into fall and is one of the common and conspicuous southern mushrooms. It may be found as far north as the Great Lakes and New England.

Microscopic features Spores 9−11 (12) × 6−8 µm, with an apical pore, dextrinoid. Cheilocystidia 27−35 (68) × 11−18 (26) µm.

Observations The specific epithet means lead colored. This species is a frequent cause of mushroom poisoning. The olive green spore deposit will serve to identify it only if the specimens are not sterile—be sure spores are present before judging the color of the "deposit."

142 *Chlorophyllum molybdites*   One-third natural size

---

143 *Lepiota sanguiflua* Murr.

Identification marks Among the species of *Lepiota* that turn yellow then some shade of red or brown where injured, this one is unusual in having the cuticle of the cap over the disc and the scales very dark brown to almost

143 *Lepiota sanguiflua*   Two-thirds natural size

black. The scales are small, not coarse. The gills are very pale yellow rather than white. The stalk is only slightly enlarged toward the base, minutely scurfy, and bears a band- or collarlike thin annulus that often disappears by maturity.

Edibility   Not reported.

When and where   Murrill found it under live oak in rich soil in Gainesville, Florida. It occurs from Florida into Texas and may be found on piles of leaves and similar accumulations of plant material in the summer.

Microscopic features   Spores 7.5−10.5 × 5.5−6.5 (7.5) µm, dextrinoid; apical pore small. Cheilocystidia 30−52 × 10−15 µm.

Observations   Murrill reported that when the stalk was cut it released a red juice, hence the specific epithet which means bloody juice.

Lepiota besseyi H. V. Smith & Weber (in press) **144**

Identification marks   Several species share the features of a scaly cap, color change to orange red or brownish red on bruised areas, and stalk that is distinctly to slightly swollen at some point. In this species the scales are small and, like the disc of the cap, cinnamon brown; the gills are white; and the stalk bears bands and patches of colored material below the annulus. The stalk is not greatly enlarged.

Edibility   Not reported but presumably edible.

When and where   Scattered to gregarious in lawns and on mulches composed of wood and bark chips; summer; known to date only from Lake Jackson, Texas, and Honolulu, Hawaii.

Microscopic features   Spores 9−11 × 7−8 µm, dextrinoid, witn an apical pore. Pleurocystidia 60−75 × 15−20 µm, abundant; cheilocystidia similar.

Observations   This species is named for E. A. Bessey (1877−1957), long a professor of botany at Michigan State University, who collected it while in Hawaii. *L. americana* is a well-known edible mushroom and closely related to *L. besseyi*. *L. americana* differs in lacking pleurocystidia and in having coarse scales on the cap, and a glabrous stalk that is enlarged in the middle.

144   *Lepiota besseyi*   One-half natural size

145　*Lepiota naucinoides*　　　　　　　　One-half natural size

145　*Lepiota naucinoides* Pk.

Identification marks　Like the death angels, these fruiting bodies are white overall and have a distinct annulus on the stalk. In the *Lepiota*, however, no volva is present at the base of the stalk although it is often slightly swollen. White is the usual color of the gills but sometimes they become pinkish gray in age.

Edibility　Not recommended; it has caused some gastrointestinal upsets and can be confused with the destroying angels if the base of the stalk is not examined carefully.

When and where　Scattered to gregarious in lawns and grassy places; summer to late fall during warm weather; widely distributed and common.

Microscopic features　Spores 8−10.5 (15) × 6−7 (8) μm, strongly dextrinoid; apical pore minute. Cheilocystidia 25−40 × 7−12 μm.

Observations　Those who can tolerate it enjoy eating this mushroom, but we know of no way to predict one's reaction in advance. This species has long been called *L. naucina* but that name cannot be used for technical reasons, so we are using Peck's name instead that means *naucina*-like.

146　*Lepiota procera* (Scopoli : Fr.) S. F. Gray (Parasol Mushroom)

Identification marks　The long narrow stalk is often the first feature to attract one's attention. It varies from only about 1.5 times as long as the diameter of the cap to 3 or 4 times as long and is often swollen at the base. At first the caps are smooth, but as they expand broad flat scales develop as the cuticle is pulled apart. The annulus has a double edge and is not pendant; it is moveable in age.

Edibility Edible and choice. Keep in mind, however, that in North America this is a variable species and some "forms" are more desirable for the table than others.

When and where Common some years east of the Great Plains during late summer and fall under hardwoods and in mixed forests where pine is present as well as under brush and in waste places. It is more abundant in the South than elsewhere in the United States.

Microscopic features Spores 14−18 × 9−12 µm, dextrinoid, with an apical pore. Cheilocystidia 20−40 × 5−12 µm.

Observations The stalk is colored like the cap and may become furfuraceous in age. This is another collective species which will no doubt be broken up into several species at a later date. *Procerus* means very tall or high.

146 *Lepiota procera*   About one-half natural size

Lepiota humei Murr.   **147**

Identification marks At maturity the cap is nearly plane and white with a pale dingy brownish yellow umbo. The scales are white or nearly so near the margin and match the umbo near the center. Note the prominent striations at the margin of the cap. The stalk is smooth, whitish to pale pinkish gray or dull reddish, equal in diameter above a small bulb, and bears a persistent white annulus with a brown, double edge. Injured areas on the cap and stalk turn maroon then dark brown without a yellow stage.

Edibility Not recommended; we have no information on it.

When and where Gregarious to scattered in open areas, lawns, gardens, oak-pine woods, and on manure piles; July to December. It was described from near Gainesville, Florida, and is to be expected in Florida and the coastal plain; often abundant.

Microscopic features Spores 7.5−9 (10.5) × 5−6 (7) µm, dextrinoid, lacking an apical pore. Cheilocystidia 25−50 × 4.5−7.5 µm.

Observations The collectors of the type collection, Dr. H. H. Hume and his son are commemorated in the specific epithet. The basidia are 2-spored, an unusual feature in this genus. Because of the striate margin of the cap this species would be a *Leucocoprinus* in some classifications.

147 *Lepiota humei*   About one-half natural size

148   *Leucocoprinus fragilissimus* (Ravenel in Berk. & Curt.) Pat.

Identification marks   Delicate, almost transparent, yellow caps are characteristic of this species. The slender stalks are fragile and often break no matter how careful one is in handling them.

Edibility   Too ephemeral and thin fleshed to be worth testing.

When and where   Solitary to scattered in mixed woods; summer; common along the Gulf of Mexico and to be expected in the coastal plain and Florida. Ravenel collected it in South Carolina; more recently it has been reported from Sri Lanka and Zaire.

Microscopic features   Spores $9-14 \times 7-8.5$ µm, dextrinoid; apical pore prominent. Pleurocystidia absent. Cheilocystidia $13-25 \times 9-15$ µm.

148 *Leucocoprinus fragilissimus*   Three-fourths natural size

Observations *Fragilissimus* means most fragile. *L. magnicystidiosus* is similar in appearance but has a browner disc on the cap and both pleurocystidia and cheilocystidia are present and large (35−105 × 16−25 μm). To date *L. magnicystidiosus* is known only from Texas and Tennessee.

Leucocoprinus luteus (Bolton) Godfrin   **149**
(Yellow Cottony Agaric)

Identification marks Shades of sulphur to lemon yellow are characteristic of all parts of the fruiting body. In contrast to *L. fragilissimus*, the fruiting bodies are fleshier and often gregarious to cespitose. The soft warts on the cap are easily brushed off.

Edibility Reported to be poisonous, but an acquaintance of ours has eaten it several times without ill effect. Until more is known about its chemical constituents, we do not recommend experimenting with this species.

When and where Gregarious to cespitose on soil or more often where there is rich organic matter such as decaying hay, mulches, compost heaps, or leaf piles; late spring to early fall; widely distributed and common in the South. It is also common in greenhouses, flower pots, and even plantings in shopping malls in the cooler parts of North America and is widely distributed in the warmer parts of the world.

Microscopic features Spores 9−10.5 × 6−7.5 μm, dextrinoid; apical pore prominent. Cheilocystidia 40−60 × 10−15 μm.

Observations This species has been called *L. birnbaumii*, but that name has been dropped because of recent changes in the rules for naming fungi. *Luteus* means deep yellow or golden yellow.

149 *Leucocoprinus luteus*   About two-thirds natural size

Leucocoprinus breviramus H. V. Smith & Weber   **150**

Identification marks Easily detached, soft, rounded warts are scattered over the surface of the snow white cap. The flesh is thin and the cap margin striate. There is a thin annulus that often disappears by maturity. The lower part of the stalk is slightly swollen, but there is no volva.

Edibility   Little is known about its edibility; we have one report that it is edible but do not recommend it unreservedly.

When and where   Singly or in small clusters on lawns, in gardens, and on wet hay; summer and fall; known from Texas and Mississippi and to be expected generally in the warmer parts of the South.

Microscopic features   Spores 7.5–9 (10) × 5.5–6.5 μm, dextrinoid, with an apical pore. Cheilocystidia 34–67 × 9–12 μm.

Observations   *Breviramus* means short branch and refers to the type of branching observed on the cells in the cap cuticle. *L. cepastipes* is a closely related species that lacks warts on the disc of the cap, typically fruits in large clusters (often on leaf piles), and whose caps become pale grayish or brown in age. It also occurs in the South and is widely distributed in eastern North America.

150   *Leucocoprinus breviramus*   About two-thirds natural size

---

151   ***Leucocoprinus longistriatus***   (Pk.) H. V. Smith & Weber

Identification marks   Striations extend from the margin nearly to the disc on these pale tan caps. Darker tan to light brown fibrils are appressed over the disc and form soft fibrillose scales elsewhere. The smooth stalk tapers upward from the base and bears a tight annulus that is often lost by maturity. Neither the odor nor the taste are remarkable.

Edibility   Too thin fleshed to be significant as an esculent; not recommended.

When and where   Summer; on the ground or on mulches, in lawns, gardens, parks, and open woods. F. S. Earle collected it in Alabama and sent specimens to Peck who described the species. It is probably common in Florida and the coastal plains and has been reported from the West Indies, but the details of its distribution remain to be learned.

Microscopic features   Spores 6–8 (9) × 4.5–5 μm, not dextrinoid; lacking a prominent apical pore. Cheilocystidia 21–33 × 8–12 μm.

Observations   *Longistriatus* means having long lines or grooves.

151 *Leucocoprinus longistriatus*   One-half natural size

Leucocoprinus lilacinogranulosus (Hennings) Locquin'   152

Identification marks   The dull purple to purplish brown disc of the cap and scales, striations on the cap, small- to medium-sized fruiting bodies, and somewhat bulbous stalk form a distinctive combination. In age the gills may be crinkled and pale pink. We noticed no major color changes on injured areas or as a result of drying.

Edibility   We have no information on it; not recommended.

When and where   Little is known about the distribution of this species in North America; these specimens were found in a longleaf pine nursery in Mississippi in July. The species was described from material collected in the Berlin Botanical Garden; there are several reports of its occurrence in greenhouses and cultivated areas in Europe.

Microscopic features   Spores 7.5−11 × 5−7.5 µm, dextrinoid; apical pore prominent. Cheilocystidia collapsed by maturity.

Observations   The specific epithet refers to the lilac granules (scales) on the cap. Old spores may be pale rose when mounted in KOH and viewed with a microscope.

152 *Leucocoprinus lilacinogranulosus*   Three-fourths natural size

Tricholomataceae

In a nutshell, this family includes those gilled mushrooms with white to lightly colored spores in deposits (pink, pinkish buff, pinkish gray, or pale yellow), gills attached to the stalk, and fruiting bodies in which the cap and stalk do not separate cleanly. In addition, the spores lack an apical pore, they may be smooth or ornamented, and, if pink, they are not angular in outline. Veils may be present or absent.

This family is an important one for mycophagists because it contains many excellent edible species and, not surprisingly, some poisonous ones. It is one of the largest families of gilled mushrooms in the world.

Key to Genera and Species

1. Stalk attached to center of cap 3
1. Stalk attached to cap at its edge or distinctly off center 2
   2. Surface of cap fuzzy, dull rosy purple to dusky rose . . . .
. . . . 153. *Panus rudis*
   2. Cap glabrous to fibrillose, white to gray or brown . . . .
. . . . (p. 180) *Pleurotus*
3. Fruiting on dead mushrooms or old magnolia cones 4
3. Not as above 5
   4. Fruiting on remains of other mushrooms . . . .
. . . . 154. *Asterophora parasitica*
   4. Fruiting on old magnolia cones . . . . 155. *Strobilurus conigenoides*
5. Fruiting bodies fleshy; gills waxy in appearance, distant, lavender, purple, or pinkish . . . . (p. 182) *Laccaria*
5. Not as above 6
   6. Base of stalk a pseudorhiza (greatly elongated and rootlike)
. . . . 156. *Oudemansiella radicata*
   6. Not as above 7
7. Fruiting body fleshy, bright orange to yellow orange overall; fruiting singly or in clusters and associated with trees, stumps, or buried wood . . . . 157. *Omphalotus illudens*
7. Not as above 8
   8. Stalk yellow when young, dark brown and velvety in age; cap slimy when fresh; fruiting on wood of hardwoods . . . .
. . . . 158. *Flammulina velutipes*
   8. Not as above 9
9. Stalk 1–3 mm thick, bright yellow, and viscid; caps yellow to greenish yellow; fruiting on wood of conifers . . . . . . 159. *Mycena epipterygia*
9. Not as above 10
   10. Veil present on young specimens, typically leaving an annulus on the stalk 11
   10. Veils absent, or if present not forming an annulus 12
11. Cap covered with appressed pinkish gray to pinkish brown or dark brown fibrillose scales and patches; stalk similarly ornamented below the annulus . . . . 160. *Armillaria caligata*
11. Cap typically with erect to ascending fibrillose scales when young, yellow, honey brown or cinnamon brown; stalk basically glabrous below the annulus . . . . (p. 184) *Armillariella*

12. Fruiting on wood or woody debris, sometimes around stumps or arising from buried wood 13
12. Not as above; fruiting on leaves, humus, compost piles, in lawns, etc. 15
13. Cap glabrous or matted fibrillose and then depressed on the disc; flesh tough and pliant .... (p. 187) *Lentinus*
13. Cap fibrillose-scaly to scaly at least when young or if glabrous fleshy; flesh firm but easily cut or broken 14
14. Fruiting in clusters with the stalk bases often confluent or touching one another; gills decurrent in age .. (p. 184) *Armillariella*
14. Fruiting in groups but not consistently clustered; gills not decurrent in age .... 161. *Tricholomopsis formosa*
15. Gills notched where they join the stalk; stalk fleshy .... (p. 188) *Tricholoma*
15. Gills broadly adnate to decurrent or rarely attached to a collar or practically free 16
16. Fruiting bodies fleshy; stalk typically breaking when bent; gills broadly adnate to decurrent .... (p. 190) *Clitocybe*
16. Not as above; fruiting bodies pliant but seldom fleshy; stalk pliant when bent; gills variously attached but not decurrent .. .. (p. 193) *Collybia* & *Marasmius*

Panus rudis Fr.   **153**

Identification marks   These tough mushrooms have fan-shaped or spathulate caps that arise directly from wood or have a short eccentric stalk. A dense covering of soft to firm hairs is present on the caps. The latter are some shade of violet to purple when young and become reddish brown to tan in age. No gelatinous tissues are present in the fruiting bodies. The spores are yellowish in deposits.

Edibility   Not poisonous according to reports but tough and not recommended.

153   *Panus rudis*   Two-thirds natural size

When and where Cespitose to gregarious on recently downed or cut wood of hardwood trees; common and widely distributed in North America. It can fruit throughout the growing season.

Microscopic features Spores 4.5−6 × 2.5−3.5 μm. Pleurocystidia 35−55 × 9−14 μm, with thickened walls. Cheilocystidia 22−45 × 8−14 μm. Cuticle of cap of interwoven to ascending thick-walled hyphae, some in fascicles.

Observations This species usually fruits on stumps and logs in exposed situations and not on wood in late stages of decay. The specific epithet means rough, *Panus* means all ear, a reference to the fact that the fruiting bodies generally consist only of a cap without a stalk.

---

154 *Asterophora parasitica* (Bull. : Fr.) Sing.

Identification marks A relatively small number of gilled mushrooms parasitize other mushrooms. Within this group, this species is distinguished by its white to light gray, silky caps which have a strongly incurved margin at first; distinct gills; and farinaceous taste. Neither a volva nor small nodules (sclerotia) within the host mushroom are present.

Edibility Of no consequence as an esculent.

When and where Clustered to scattered on the fruiting bodies of other mushrooms, particularly members of the Russulaceae; late summer and fall; widely distributed in North America but uncommon outside the southern Appalachians. It has been reported from Eurasia, Africa, and Papua New Guinea as well.

Microscopic features Spores formed on basidia few, 5−6 × 3−4 μm, smooth; chlamydospores (formed in series in the hymenium) 12−16 × 9−11 μm, thick walled, smooth. Cuticle of cap of appressed radially arranged hyphae.

Observations Each chlamydospore develops inside a broadly fusoid cell; the final product resembles a fat sausage in a long casing. *A. lycoperdoides* also fruits on old mushrooms. Its cap is covered with brown, globose, spiny chlamydospores so it looks powdery; it often lacks gills. The genus name means star spore; *parasiticus* means parasitic.

154 *Asterophora parasitica*   About natural size

Strobilurus conigenoides (Ellis) Sing. **155**

Identification marks This tiny mushroom fruits, and presumably grows, on old magnolia cones and inflorescences. The cap is tan in the center and lighter at the margin to whitish or pale buff overall and seldom over 1.5 cm broad. The stalk is slender and almost hairlike. Where it joins the cone there is a cottony mass of buff to light brown mycelium.

Edibility Too small to consider.

When and where Late summer through fall on old magnolia cones; it was described from specimens collected in southern New Jersey, and is widely distributed and common in the South.

Microscopic features Spores 4.5–6 × 2–3 µm, smooth. Pleurocystidia (25) 40–52 × 9–14 µm, walls often thickened. Cuticle of cap a hymeniform layer with projecting pileocystidia scattered through it.

Observations This is a characteristic mushroom in the South and can be found wherever old magnolia cones are allowed to accumulate and decay. *Strobilus* is a Latin word for a cone; the specific epithet means resembling the species once known as *Collybia conigena*.

155 *Strobilurus conigenoides*   Natural size

Oudemansiella radicata (Relhan : Fr.) Sing. **156**

Identification marks The combination of the long rooting base (pseudo-rhiza), widely spaced attached gills, and lack of veils is important. The pseudo-dorhiza may extend at least a foot into the ground. In color, texture, and size the cap varies greatly. It may be only 2–3 cm or up to 10 cm broad, white or more often some shade of brown, and distinctly slimy when wet to dry to the touch. The spores are white in deposits.

Edibility Edible; the caps alone are usually eaten because the stalks are inclined to be fibrous.

When and where Solitary to gregarious on, arising from, or near well-decayed stumps or roots of hardwood; early spring to fall; common and widely distributed in eastern North America. We have collected it when morels were fruiting in late March.

Microscopic features Spores (14) 15–18 × 9–12 µm, inamyloid. Pleurocystidia 70–135 × 18–30 µm, rare. Cheilocystidia 45–80 × 12–15 µm. Cuticle of cap a hymeniform layer that thickens by intrusion of new cells as the specimens age; viscid.

Observations C. A. J. A. Oudemans (1825–1906), for whom the genus is named, was a Dutch physician and botanist. *Radicatus* means having a root or roots. The "root" of course is a pseudorhiza or false root, not a true one.

156 *Oudemansiella radicata*   About one-half natural size

---

157 *Omphalotus illudens* (Schw.) Bresinsky & Besl (Jack-O'-Lantern Mushroom)

Identification marks These mushrooms arise singly or more often in dense clusters at the base of living trees as well as from stumps and buried wood. All parts are some shade of orange to yellow orange. The cap is smooth; the gills are decurrent, thin, crowded, and sharp edged; and often the stalks are fused at the base.

Edibility Poisonous. Most people who eat it experience violent nausea and vomiting.

When and where Late summer into fall or until the onset of cold weather; common and widely distributed in eastern North America but most abundant in the South. Schweinitz found it in North Carolina.

Microscopic features Spores 4–5 × 3.5–4.5 µm, smooth. Hymenial cystidia not observed. Cuticle of cap a thin ixocutis.

Observations The gills of actively growing specimens will glow in the dark, hence the common name. It is sometimes confused with *Cantharellus cibarius* or *Armillariella tabescens* much to the regret of the mycophagist, but attention to detail will suffice to avoid this mistake. Anyone who believes that all mushrooms growing on stumps are edible probably has not eaten this species or its relatives. *Omphalotus* is derived from words meaning naval and ear; *illudens* means deceiving.

157 *Omphalotus illudens*   One-half natural size

Flammulina velutipes (Wm. Curtis : Fr.) Sing.   **158**
(Velvet Foot)

Identification marks In wet weather the caps are slimy, in dry weather they appear glazed. No veils are present. Young stalks are striate, glabrous, and dull orange brown to yellow brown. As they mature a rich brown to blackish brown pubescence develops from the base upward.

Edibility Edible when cooked; popular in the Orient where it is cultivated commercially.

When and where In dense clusters on logs, stumps, and standing dead trees, especially elm, aspen, and willow; widely distributed in North America. It is one of the few edible mushrooms that fruits from late fall to early spring during wet cool weather.

Microscopic features Spores 6–8 × 3–4 μm. Cheilocystidia 30–45 × 8–15 μm, abundant. Clamp connections present. Cuticle of cap an ixotrichodermium to ixolattice with scattered cells 52–120 × 6–14 μm that have light rusty brown walls.

Observations The specific epithet means velvet foot or stalk. *Flammulina* means pertaining to *Flammula*. Because it often fruits during mild, damp periods in the winter, this species is also known as the winter mushroom. Although it occurs in the South, it is seldom abundant, at least at low elevations.

158 *Flammulina velutipes*   About three-fourths natural size

159 *Mycena epipterygia* var. *caespitosa*   About natural size

159   *Mycena epipterygia* var. *caespitosa* Thrs.

Identification marks   This variety fruits in a gregarious to cespitose manner on or adjacent to rotting pine logs and stumps. Both the cap and stalk are slippery from a layer of slime. The caps are small, usually less than

2 cm broad, grayish olive in the center to bright yellow at the margin. The bright yellow to orange yellow stalk has stiff hairs at the base. Both the odor and taste are somewhat pungent and farinaceous to bleachlike.

Edibility Of no consequence as an esculent.

When and where In clusters on rotting pine logs and stumps; November to early January along the Gulf of Mexico. It was described from near Richards, Texas, and is probably widely distributed in the coastal plain and Florida.

Microscopic features Spores 7.5–11 × 5.5–8 µm, amyloid. Cheilocystidia 30–38 × 6–9 µm, embedded in the gelatinous gill edge and bearing antlerlike or twisted, fingerlike projections. Cuticle of cap an ixolattice.

Observations *M. viscosa* is similar in color, fruits in troops on pine needles, and has stockier fruiting bodies. *M. epipterygia* var. *epipterygia* fruits on the ground under conifers and is widely distributed. The specific epithet means a little wing; *caespitosus* means growing in tufts.

Armillaria caligata Viviani   **160**
(Booted Armillaria)

Identification marks The outer veil is dull to dark cinnamon brown, breaks up into flat patches on the cap, and leaves a membranous annulus and series of zones of colored fibrils on the stalk. The stalk typically tapers to a somewhat pointed base. The gills are white, close, and only moderately broad. The type variety, var. *caligata*, has a mild odor and taste, var. *nardosmia*, shown here, has a pungent to disagreeable odor and taste.

Edibility Avoid the bitter varieties and use the mild ones for eating. They are reported to be edible and good. The matsutake of Japan is a related species that is a popular esculent.

When and where Scattered to gregarious under oaks particularly where blueberries, huckleberries, and their relatives grow. The heaviest fruitings east of the Great Plains occur in the fall.

Microscopic features Spores 6–7.5 × 4–5 µm, inamyloid. Hymenial cystidia absent. Cuticle of cap of appressed hyphae.

Observations The specific epithet refers to a soldier's boot, probably the sheath that "boots" the base of the stalk. *Nardosmia* means having the odor of nard. *Amanita hesleri* might be confused with this species but, among other differences, it has amyloid spores.

160   *Armillaria caligata* var. *nardosmia*   About one-half natural size

161 *Tricholomopsis formosa*

Natural size

161 *Tricholomopsis formosa* (Murr.) Sing.

Identification marks Ascending to recurved rusty brown to tawny scales cover the cap; similar colors also occur on the stalk but the gills are whitish to cream color. There is no veil or annulus. Both the odor and taste were described as strongly disagreeable and earthy but they are often milder. Fruiting bodies often occur in clusters, but each stalk is distinct.

Edibility We have no information on it; not recommended.

When and where On dead wood such as logs, stumps, piles of slash, and sometimes arising from buried roots or on sawdust; summer into late fall. Little is on record about its distribution; it is to be expected at least from North Carolina south into Florida and west into Texas. It was described from Florida.

Microscopic features Spores 5–7 × 4.5–6 µm, inamyloid. Cheilocystidia 37–60 (70) × 7.5–15 µm. Cuticle of cap of hyphae grouped into fibrils. Clamp connections present.

Observations The brown colors and scaly cap may suggest *Pholiota* but the spores are white and there is no veil. *Formosus* means handsome or beautiful. *T. rutilans* also occurs in the South, has rosy red to maroon fibrils over a yellowish ground color on the cap, and fruits on old pine logs.

Pleurotus

Members of this genus fruit on wood; either lack a stalk or have an eccentric to lateral one; have white to pale lilac gray spores in deposits; and fleshy fruiting bodies. They are effective decayers of wood and sawdust. Some have been cultivated on straw. The genus name means side ear or cap and presumably refers to the position of the stalk.

Key to Species

1. Veil present on young specimens; injured areas staining yellow . .
. . . . 162. *Pleurotus dryinus*
1. Veil absent; injured areas not staining as above . . . . . . 163. *P. ostreatus*

Pleurotus dryinus (Pers. : Fr.) Kumm.   **162**

Identification marks   Large, white, softly floccose to scaly, tough caps that do not decay rapidly are characteristic of this species. The scales may become dingy yellow brown in age. A thin veil is present in buttons; at maturity a thin line of matted fibrils on the stalk and occasionally a few on the edge of the cap are all that remain. The gills are subdistant and at most only moderately decurrent—unlike the oyster mushroom. In addition there is a dense velvety tomentum on the stalk, the odor is strong and fragrant, and the spore deposits are white.

Edibility   Nonpoisonous; use young caps as old ones are likely to be tough.

When and where   Solitary to scattered, or sometimes in clusters on stumps, logs, and trunks of living or dead hardwoods, rarely on conifers; late fall to spring along the Gulf Coast, summer northward; widely distributed but seldom common.

Microscopic features   Spores 9–12 (17) × 3.5–5 µm. Cuticle of cap of loosely interwoven hyphae.

Observations   Few species of pleurotoid mushrooms with white spores have a veil. Yellow stains may develop on the fruiting bodies in age. The specific epithet refers to oak, a common substrate.

162   *Pleurotus dryinus*   About two-thirds natural size

163 *Pleurotus ostreatus* (Jacquin : Fr.) Kumm.
(Oyster Mushroom)

Identification marks Important features are the smooth caps that are white to gray or blackish brown; decurrent, broad, whitish gills; and the stalk that, when present, is attached at or toward the side of the cap (lateral to eccentric). Spore deposits are white or lilac gray. The flesh is firm and meaty.

Edibility Edible and popular; the common name supposedly refers to the shape of the cap, not the flavor. A sharp tap on the top of a cap, held with the gills toward the ground, usually will dislodge the small beetles that often live between the gills.

When and where Clustered, scattered, or in shelving masses on a variety of hardwoods, both living and dead; widely distributed and common. It fruits from fall into spring in the South, and spring into fall in the North. In Mississippi we have seen where it could be collected by the bushel in bottomland hardwood forests on willows and oaks.

Microscopic features Spores $9-11 \times 3-4$ µm. Cuticle of cap a compact zone of radially arranged hyphae.

Observations Successive crops can appear on a given log or tree and provide fresh specimens several times during the fruiting season. Oyster mushrooms are cultivated commercially in some areas. This is another collective species that includes several variants, none known to be poisonous.

163 *Pleurotus ostreatus*   About one-half natural size

Laccaria

Members of this genus have violet, lilac, purple, or rose to deep flesh pink gills that long retain their color and are thick and "waxy" in appearance; caps that are dry and often become furfuraceous to squamulose in age; fibrous stalks; and white to pale lilac spores in deposits. The spores are inamyloid. No veils are present.

These mushrooms fruit on the ground in a variety of habitats, particularly on bare or poor soil. Sometimes they are mistaken for members of the Hygrophoraceae with similar gills. The name *Laccaria* is said to refer to milk or lacquer. Its relevance to the species currently placed in the genus is, at best, obscure.

Key to Species

1. Fruiting in sand, especially on dunes and beaches; base of stalk deeply inserted into the sand and covered with sand . . . . 164. *Laccaria trullisata*
1. Fruiting in thin woods and their borders; base of stalk not deeply inserted into the substrate . . . . 165. *L. ochropurpurea*

Laccaria trullisata (Ellis) Pk.   **164**

Identification marks   This is one of a small group of species that typically fruits on barren sandy soil and sparsely vegetated beaches and dunes. The gills are purplish to violaceous, waxy, and somewhat distant. At least half, often more, of the stalk is covered by sand and the caps may not be elevated much above the sand.

Edibility   Edible but dangerous to dental work and teeth because of the adhering sand.

When and where   Scattered to cespitose in sand and barren areas after heavy rains in the late summer until the end of the growing season; widely distributed on barrier islands, undisturbed beaches (if such exist) and dunes along the Atlantic and Gulf coasts and the Great Lakes. It was described from a sandy field in southern New Jersey.

Microscopic features   Spores (16) 18−23 × 6−8 µm, smooth to minutely roughened.

Observations   The specific epithet refers to a bricklayer's trowel, a reference to the shape of the stalk.

164   *Laccaria trullisata*   Three-fourths natural size

165 *Laccaria ochropurpurea*   One-half natural size

---

165   *Laccaria ochropurpurea* (Berk.) Pk.

Identification marks   When young and fresh the gills are purple and both the cap and stalk are purplish brown. In age and as moisture is lost the cap and stalk become whitish to grayish, or yellowish tan and the surface of the stalk breaks up into scales. The gills, however, retain their color.

Edibility   Edible and enjoyed by some mycophagists, not others. Fresh specimens have a disagreeable taste that is lost during cooking.

When and where   Scattered to gregarious on the ground in thin deciduous woods and woods borders; summer until the onset of cold weather. It was described from collections made in Ohio, but the species is widely distributed and often common east of the Great Plains generally.

Microscopic features   Spores 8–10 × 7.5–9.5 µm, echinulate.

Observations   The specific epithet means ocher and purple. Fruiting bodies of this species are the largest known for North American members of this genus. The waxy gills and pale lilac spores in deposit distinguish it from *Clitocybe nuda* and species of *Cortinarius* with blue to purple gills.

Armillariella

This genus is distinguished by the combination of adnate to short-decurrent, white to whitish gills; white to light cream spores in deposits; fleshy fruiting bodies with pale flesh, and central stalks; and habit of fruiting in clusters on or arising from wood. Fibrillose scales are usually present on the cap and black strands of hyphae (rhizomorphs) are often associated with the fruiting bodies.

Two species are included here, but it should be understood that they are better thought of as species complexes that include sev-

184

eral variants in each "species." At present we cannot place each variant into a species or variety with confidence; furthermore, many of the characters used to distinguish them are not suited to a field guide approach. Both species treated here have long been considered safe and are popular edible mushrooms, but there is an increasing number of reports of gastrointestinal upsets following ingestion of these mushrooms, even when the specimens were thoroughly cooked. The name of the genus is the diminutive of *Armillaria* which means a ring. Some authors are now using the name *Armillaria* for this genus.

Key to Species

1. Partial veil present and leaving a cottony to submembranous annulus on the stalk . . . . 166. *Armillariella mellea*
1. Partial veil and, thus, annulus absent . . . . 167. *A. tabescens*

Armillariella mellea (Vahl : Fr.) Karst. **166**
(Honey Mushroom)

Identification marks Important features include the white to cream-colored spore deposits, whitish gills and flesh, cottony veil that forms the membranous to cottony ring, and conspicuous to minute scales on the caps. The caps vary from yellow to honey brown or cinnamon brown and are often slimy, particularly in wet weather.

Edibility Cooked specimens can be eaten and enjoyed by most people. This species has caused severe indigestion when eaten raw, with alcoholic beverages, and occasionally when cooked. The slime in the cap may be pronounced in cooked and pickled material. Thorough cooking and moderation are advised when sampling this species.

166 *Armillariella mellea*   Three-fourths natural size

When and where   Solitary or more often in clusters, attached to or arising from wood (either living or dead); late summer into winter in the South; widely distributed and abundant.

Microscopic features   Spores 6−9 × 4.5−6 µm. Cheilocystidia 18−27 × 9−10 µm, soon collapsing. Cuticle of cap of gelatinous, interwoven to radially arranged hyphae and hyphal fascicles.

Observations   We have not tried to distinguish the component taxa in this species complex as such determinations often rest on technical characters. In one or more of its variations this is a popular edible mushroom which is also known as the stumpy or cinnamon top. *Melleus* means honey colored.

---

167   *Armillariella tabescens* (Scopoli) Sing.

Identification marks   Like the honey mushroom this species has white spores in deposits, whitish gills and flesh, and yellow to honey brown or dull rusty brown caps decorated with fine fibrillose scales. It differs in lacking a partial veil and in having a dry cap and more slender fruiting bodies. Its fruiting bodies often form denser and larger clusters than do those of the honey mushroom.

Edibility   Edible for most people when thoroughly cooked; but it has caused cases of severe indigestion; do not indulge heavily.

When and where   In clusters attached to or arising from living trees or dead wood, especially stumps and buried roots; summer into fall; common in the South and occasional as far north as New England and the Great Lakes region. In Florida it reportedly fruits in the summer and fall and the honey mushroom fruits in the winter.

Microscopic features   Spores 6−8 (9) × 4.5−6 µm. Hymenial cystidia not observed. Cuticle of cap of appressed to loosely arranged hyphae grouped into fascicles.

Observations   *Tabescens* means wasting away or stunted. This is another species complex in which recognition of separate taxa is beyond the scope of a field guide. Compare it with the jack-o'-lantern mushroom (*Omphalotus illudens*), which has clustered fruiting bodies but decurrent orange gills and orange flesh and is poisonous.

167   *Armillariella tabescens*   One-third natural size

Lentinus

The tough consistency of the fruiting bodies and habit of fruiting on wood are important features. A stalk is present on most fruiting bodies, and the cap may be smooth or fibrillose to fibrillose-scaly. A veil is present in some species. A satisfactory classification for *Lentinus* and related genera has yet to come to our attention. The grouping used here is primarily one of convenience, the relationships of *L. detonsus* in particular being a matter of uncertainty. The name *Lentinus* means tough or pliant.

Key to Species

1. Cap fibrillose, deeply depressed over the disc; gills crowded, decurrent .... 168. *Lentinus crinitis*
1. Cap glabrous or with evanescent veil remnants near the margin, convex to plane; gills adnate, receding in age or when dried .... 169. *L. detonsus*

Lentinus crinitis (L. : Fr.) Fr.   **168**

Identification marks   These tough, thin-fleshed mushrooms fruit on dead wood. The cap is funnel-shaped to deeply depressed over the disc and is covered with dense, shiny brown fibrils and fibrillose scales. The gills are extremely crowded, decurrent, and off-white.

Edibility   Not recommended but in fairness we should report that some Indians in Brazil are reported to boil and eat it.

When and where   Scattered to gregarious on dead wood; summer; widely distributed in the tropical and subtropical regions of the world and not uncommon in the South. It was described from Jamaica.

Microscopic features   Spores 6–8 × 3 µm, absent or difficult to locate. Hymenial cystidia absent. Cuticle of cap of radially arranged thick-walled hyphae.

Observations   There is considerable controversy over the correct placement of this species—it has been placed in *Panus* as well. It is common on exposed logs and trunks around the Gulf of Mexico. The specific epithet means having tufts of long weak hairs.

168   *Lentinus crinitis*   About one-half natural size

Identification marks At first glance this appears to be a *Collybia* fruiting on a tree. The stalk is tough and eccentric. The gills are extremely crowded, stain reddish then brown when bruised, and although they meet the stalk in fresh specimens, they recede during drying. The taste is reminiscent of garlic.

Edibility Not recommended; too little has been published about its edibility to allow for generalizations.

When and where Gregarious to scattered on trunks of living or dead trees and branches; summer. It occurs in the South and in Central America, the West Indies, and around the Gulf of Mexico.

Microscopic features Spores 4.5–6 × 2–3 μm, smooth, inamyloid. Cystidia absent from the hymenium. Cuticle of cap of matted, radiating to interwoven thick-walled hyphae.

Observations The presence of thick-walled hyphae in the cap accounts for the toughness of the fruiting bodies and is one reason why this species is not placed in *Collybia*. Our identification is a "best guess," there is little information available on this group of species. Earle, in 1909, described a new genus *Lentinula* for this and similar species. Whatever one chooses to call it, it is a common species around the Gulf of Mexico and in the West Indies. *Detonsus* means to shear off.

169 *Lentinus detonsus*   About one-half natural size

Tricholoma

The notched gills, fleshy fruiting bodies, terrestrial habit, and smooth, inamyloid, and white or whitish spores are important characters. Veils are present in a few species. Members of this genus fruit primarily in the late summer and fall.

Although some species are edible, others can cause serious poisonings. Anyone who wants to eat these mushrooms should first make an intensive study of the genus. The name means hairy fringe.

Key to Species

1. Cap, gills, and stalk bright greenish yellow when young; cap slimy; taste mild to farinaceous .... 170. *Tricholoma flavovirens*
1. Cap and stalk white to light gray, gills light gray; cap moist; taste acrid .... 171. *T. acre*

Tricholoma flavovirens (Pers. : Fr.) Lundell   **170**
(Man-on-a-Horse)

Identification marks The cap is plane to slightly convex but never pointed. A thin layer of slime causes dirt and debris to stick to it. The cap, gills, and stalk are basically bright greenish yellow; however, the cap may be washed with tan in the center. Although the odor is mild, the taste is farinaceous, i.e., like freshly ground wheat.

Edibility Edible; however, the adhering sand and dirt is difficult to remove completely.

When and where Scattered or in arcs, usually in sandy soil under pine; late fall into winter; to be expected throughout the piney woods of the South, widely distributed in North America; often abundant.

Microscopic features Spores 6−7.5 × 4−5 µm. Cystidia not observed. Cuticle of cap a thick ixotrichodermium that collapses in age.

Observations The specific epithet means yellow green. Another name formerly used for the species is *T. equestre*. That specific epithet means belonging to a horseman; by extension it was interpreted to mean of distinguished appearance and is the apparent source of the common name.

170   *Tricholoma flavovirens*   One-half natural size

Tricholoma acre Pk.   **171**
(Acrid Tricholoma)

Identification marks Important features of this species are the streaked, innately fibrillose, dry, gray cap that is rounded to plane in outline at maturity; the notched gills that are white at first and grayish in age; the whitish to pale gray fibrous stalk; and the decidedly acrid taste. The odor is mild.

171 *Tricholoma acre*   One-half natural size

Edibility Unpalatable; whether it is toxic is apparently not known.

When and where Gregarious to scattered in woods of broad-leaved trees; fall to early winter; widely distributed. The type collection is from Massachusetts; the species occurs as far south as Florida and in the Rocky Mountains.

Microscopic features Spores 6–7.5 × 5–7 μm. Pleurocystidia absent. Cheilocystidia 30–54 × 5–12 μm, in patches, walls light brown in KOH. Cuticle of cap of dry, radially arranged hyphae.

Observations The specific epithet means acrid. Among the species of *Tricholoma* with gray to brownish gray, innately fibrillose to fibrillose-scaly caps and white to gray gills there are both dangerously poisonous and edible species. The group is best avoided by beginning mycophagists.

Clitocybe

The fruiting body is fleshy, the gills decurrent in most species but broadly adnate in some, and veils are absent. The spores are white, pale yellow, pinkish buff, or grayish pink in deposits but are colorless under the microscope and are not angular. The margin on young caps is usually incurved to inrolled.

Several species are edible, others are poisonous and often difficult to identify. The name of the genus means sloping head or cap.

Key to Species

1. Gills decurrent; cap and stalk light to dark grayish brown; stalk enlarged toward the base .... 172. *Clitocybe clavipes*
1. Not as above 2
    2. Cap dull lilac to lavender or pinkish tan; gills dull lilac ....
        .... 173. *C. nuda*
    2. Cap white to grayish white; gills whitish .... 174. *C. dealbata*

Clitocybe clavipes (Pers. : Fr.) Kumm.   **172**
(Clubfooted Clitocybe)

Identification marks   Dull gray to gray brown tones are characteristic of the cap and stalk; the gills are white or nearly so, subdistant, and decurrent; and the stalk tapers from a large base to a much narrower apex. Spore prints are white. The odor is often fragrant, and the taste is mild.

Edibility   Not recommended; when consumption of the mushroom is followed by alcoholic beverages, symptoms similar to those of coprine poisoning (see p. 10) may develop.

When and where   Scattered to gregarious and often abundant under conifers, especially white pines and in pine plantations, but occasionally under hardwoods; summer and fall; widely distributed in eastern North America and common in the Appalachian uplands.

Microscopic features   Spores 6–8 × 4–5 µm. Cuticle of cap of radially arranged dry hyphae.

Observations   The specific epithet means clubfooted, an allusion to the enlarged base of the stalk of mature specimens.

172   *Clitocybe clavipes*   About two-thirds natural size

Clitocybe nuda (Bull. : Fr.) Bigelow & Sm.   **173**
(Blewits)

Identification marks   When young and fresh these characteristically squat, fleshy fruiting bodies are some shade of lavender to violet all over. As they age they fade to lilac tan or pinkish tan. Fresh caps may appear watery and feel slippery to the touch. No veils are present. The spore deposit color is pinkish tan to dull reddish gray. Both the odor and taste are mild.

Edibility   Edible and highly esteemed by most mycophagists. In our experience old specimens, or those not cooked within a day or so of picking, often have an unpleasant flavor.

When and where   Scattered to gregarious, sometimes in arcs, on accumulations of dead leaves and humus in forests and on leaf piles and compost heaps; widely distributed and common in North America. It fruits in late summer into the fall or winter during cool wet weather.

Microscopic features   Spores 6–8 × 3–5 μm, minutely roughened, inamyloid. Cystidia not observed on the gills. Cuticle of cap a thin ixocutis.

Observations   The colors change considerably from youth to age, but some hint of lilac persists on some part of the fruiting body. Some species of *Cortinarius* are similar in stature and coloring, but they have a thin fibrillose veil at least in buttons and rusty brown spores. Such species should be avoided by the mycophagist. *Nudus* means bare or naked.

173   *Clitocybe nuda*   About two-thirds natural size

174   *Clitocybe dealbata* (Sowerby : Fr.) Kumm. (The Sweater)

Identification marks   Dull gray lines or cracks arranged in a more or less concentric pattern on a relatively flat white to grayish white cap are characteristic of this species. The gills are crowded and often slightly decurrent. The stalk is seldom longer than the width of the cap, but is tough, fibrous, and whitish. The spores are white in deposit. Both the odor and taste are mild to pleasant.

Edibility   Poisonous; if more than a few caps are eaten, the characteristic signs of muscarine poisoning (see p. 10) usually develop. However, reports that it is deadly poisonous are erroneous.

When and where   In lawns, pastures, and open grassy areas generally, often with Scotch bonnets (*Marasmius oreades*) and occasionally in woods; summer and fall; widely distributed.

Microscopic features   Spores 4–6 × 2–3 μm. Cuticle of cap of loosely interwoven dry hyphae with scattered free tips.

Observations   Most poisonings by this species occur in children in the "grazing" stage or adults who gathered it along with Scotch bonnets, *Marasmius oreades*. The latter typically has a stalk longer than the width of the cap and rather broad distant gills. *Dealbatus* means whitewashed.

174 *Clitocybe dealbata*   Natural size

Collybia & Marasmius

Without checking the structure of the cap cuticle it is frequently difficult, if not impossible, to separate these two genera. Both have
thin-fleshed fruiting bodies (occasionally fleshy in *Collybia*) many
of which can dry out, be remoistened, and produce spores a second or third time. In *Collybia* the cuticle of the cap is of appressed
filamentose hyphae; in *Marasmius* it consists of a hymeniform layer
of rounded to clavate cells often with specialized branches on
them. *Collybia* means a small coin; *Marasmius* as a name has its
origins in a term meaning to become dry.

Key to Species

1. Cap and gills lavender to dull purple; odor somewhat like garlic ..
   .... 175. *Collybia iocephala*
1. Not as above 2
   2. Cap bright rusty orange; stalk slender, resembling a horsehair
      .... 176. *Marasmius fulvoferrugineus*
   2. Not as above 3
3. Cap the color of cream to tan or rarely light brown; gills subdistant
   and broad; typically fruiting in arcs and rings in lawns and other
   grassy areas .... 177. *Marasmius oreades*
3. Not as above 4
   4. Cap light to medium yellow brown; stalk paler than cap; fruiting
      bodies often cespitose .... 178. *Collybia dryophila*
   4. Cap pale pinkish tan to pinkish buff or almost white; stalk about
      the same color or paler; fruiting bodies scattered or in small
      clusters .... 179. *Marasmius nigrodiscus*

175 *Collybia iocephala* (Berk. & Curt.) Sing.

Identification marks   Violet to grayish violet or lilac colors are characteristic of the striate, thin, and pliant caps. The stalk is tomentose to strigose and whitish to dingy pale gray. Both the odor and taste are distinctive, variously described as garliclike, similar to old cabbage, or strong and offensive.

Edibility   Some mycophagists have suggested using it as a seasoning to replace garlic, others are repelled by the idea. Not recommended for use in quantity.

When and where   Scattered to gregarious or cespitose on humus, leaf litter, and compost heaps; summer and fall; common. It occurs from southern New England south into Florida and west to the Great Plains.

Microscopic features   Spores 6.5–8 × 2–3.5 μm, inamyloid. Cheilocystidia inconspicuous, reported to be 18–24 × 4–5.5 μm. Cuticle of cap a cutis of narrow hyphae. Clamp connections present.

Observations   The thin membranous flesh is similar to that of many species of *Marasmius* but the structure of the cap cuticle is that of a *Collybia*. When touched with potassium hydroxide (KOH) the surface of the cap becomes greenish blue. *Iocephalus* refers to the violet cap.

175   *Collybia iocephala*

Slightly larger than natural size

176 *Marasmius fulvoferrugineus* Gilliam

Identification marks   The cap is dry, ribbed, and rusty brown to brownish orange. As is typical of the genus, the flesh is very thin and pliant but tough. The yellowish white gills are widely spaced. The tough stalk is shiny, glabrous, and resembles a coarse horsehair. Both the odor and taste are mildly farinaceous.

Edibility Of no consequence; we have no data on it.

When and where Scattered to gregarious on leaf litter in mixed woods; summer and fall. It occurs in the Southeast north to New Jersey and west at least into Mississippi.

Microscopic features Spores 15–18 × 3–4.5 μm. Pleurocystidia typically absent. Cheilocystidia 8–18 × 8–12 μm, clavate to obovate with numerous fingerlike projections on them (these cells are aptly termed *broom cells*). Cuticle of cap a hymeniform layer of broom cells.

Observations *M. siccus* is similar but has a brighter orange cap; large abundant cystidia on the gills, and longer spores (16–21 × 2.8–4.2 μm). *M. fulvoferrugineus* is more common in the South, *M. siccus* in the North. *Fulvus* means tawny, *ferrugineus* means rusty.

176   *Marasmius fulvoferrugineus*   Slightly larger than natural size

---

Marasmius oreades (Bolton : Fr.) Fr.   **177**
(Scotch Bonnets)

Identification marks The combination of a white spore deposit, slender naked stalk, pallid gills which are broad and spaced relatively far apart (subdistant), naked often umbonate cap, and habit of fruiting in lawns and grassy places should distinguish it. The color of the cap varies from reddish tan to pale tan or pale buff. The fruiting bodies are slow to decay with the result that the pigment may leach out of the caps which may become nearly white.

Edibility Edible and choice; it has been a popular species for years. But beware of small white to grayish white fruiting bodies of *Clitocybe dealbata* that often appear *in* the clusters of *M. oreades*; the *Clitocybe* is poisonous.

When and where On grassy areas such as lawns, golf courses, and old fields during warm wet weather; it is common in the North but seems to be much less abundant in much of the South. It often fruits in rings or arcs and thus also is known as the fairy ring mushroom.

Microscopic features Spores 7–9 × 4–5 µm, inamyloid. Pleurocystidia absent. Cheilocystidia 20–34 × 3–5 µm. Cuticle of cap a hymeniform layer.

Observations Care should be taken *not* to collect for the table specimens from lawns or other areas that have been treated with chemicals. There is the possibility that the chemicals will be absorbed by the mycelium and passed on to the fruiting bodies. The specific epithet means of the mountains—not exactly an appropriate specific epithet for our common golf course mushroom!

177   *Marasmius oreades*   About natural size

---

178   *Collybia dryophila* (Bull. : Fr.) Kumm.

Identification marks This rather nondescript mushroom is characterized as much by what it lacks—veil, strong odor and taste, rooting stalk, bright colors—as by its positive attributes. The smooth caps have a greasy feel when moist; the gills are close and white to pale yellow; the stalk is whitish to yellowish or light reddish brown near the base; and the spores are white in deposit.

Edibility Edible for many people, but others experience gastrointestinal upsets after eating it; try with caution if at all.

When and where Seemingly present in every upland hardwood forest in eastern North America in the summer and fall; it often fruits in clusters.

Microscopic features Spores 4.5–6 (7) × 3–3.5 µm. Hymenial cystidia absent.

Observations Fruiting bodies infected with *Christensenia mycetophila*, another fungus, develop irregular, rounded, lobed growths on them or are completely distorted. Such fruiting bodies may occur with normal specimens. *Dryophilus* means oak loving.

Marasmius nigrodiscus (Pk.) Halling

Identification marks   For a *Marasmius*, this is a large mushroom; the caps may be up to 11 cm broad. At first the caps are light brown to dull yellowish brown but they fade to pale buff or ivory. When moist and fresh, they are translucent striate but they become opaque as they fade. Neither the odor nor the taste are strong. The stalk is dry, chalky, and often striate.

Edibility   We have heard a rumor that it is edible, but have no other information on it.

When and where   Scattered on the ground under hardwoods or in mixed woods; summer and early fall; widely distributed in eastern North America and especially abundant south of the Ohio River; the limits of its range remain to be learned.

Microscopic features   Spores 6–7 × 3–4.5 µm, inamyloid. Pleurocystidia 60–105 × 9–13 µm, abundant; cheilocystidia similar but smaller. Cuticle of cap a hymeniform layer of clavate to pyriform cells.

Observations   This species represents a complex group of mushrooms in need of detailed study that is well represented in the South. *Nigrodiscus* means dark or black disc.

178   *Collybia dryophila*   Slightly less than natural size

179   *Marasmius nigrodiscus*   One-half natural size

Pluteaceae

The combination of free gills, relatively soft fruiting bodies, cap and stalk that separate easily (and fit together like a ball and socket), and dusky rose to dull red spore deposits is diagnostic for this family. The spores are smooth and lack an apical pore. In the gills, the hyphae are arranged in a distinctive manner.

Species of *Pluteus* typically fruit on wood and lack a volva. Mycophagists generally give them a low rating because of the soft texture and habit of usually fruiting in small quantities. *Pluteus* means a shed or penthouse and refers to the shape of the cap.

Species of *Volvariella* can be distinguished from those of *Pluteus* by the presence of a distinct volva. They fruit on wood, soil, and accumulations of organic debris. Several species are edible and good when young but a few have a poor reputation. In North America this genus reaches its greatest diversity in the South. *Volvariella* is the dimunitive of *Volvaria*.

Key to Species

1. Volva present 2
1. Volva absent 3
    2. Cap white to yellow and finely silky fibrillose to scaly; fruiting on wood or from wounds on trees . . . . 180. *Volvariella bombycina*
    2. Cap blackish to gray, appressed fibrillose; fruiting on piles of leaves, old straw, etc. . . . . 181. *Volvariella volvacea*
3. Cap gray to grayish brown . . . . 182. *Pluteus cervinus*
3. Cap white overall or only the center tinged brown to gray . . . .
    . . . . 183. *Pluteus pellitus*

---

180   *Volvariella bombycina* (Schff. : Fr.) Sing.

Identification marks Its habit of fruiting on wood, presence of a persistent membranous volva, and white cap covered with silky fibrils, form a distinctive combination of characters. The gills are free and dull pink at maturity. There is no annulus.

Edibility Edible and good.

When and where Solitary to gregarious, often growing from wounds on living trees as well as on dead trees and logs of hardwoods, particularly elm, maple, magnolia, beech, and water tupelo; summer and fall during hot weather. It is widely distributed in eastern North America but seldom abundant.

Microscopic features Spores 7−9 × 5−6 μm. Pleurocystidia and cheilocystidia (26) 60−105 (122) × (8) 15−35 (57) μm. Cuticle of cap of fascicles of hyphae.

Observations *Bombycinus* means silky. This distinctive species is easily identified but not as readily found—one may not see it for several years then find it repeatedly. Variety *flaviceps* has a bright yellow cap but is similar in other respects to var. *bombycina*. Variety *flaviceps* has been reported from Florida where it fruited on magnolias.

180 *Volvariella bombycina*   One-half natural size

Volvariella volvacea (Bull. : Fr.) Sing.   **181**
(Straw Mushroom)

Identification marks At maturity the gills are dull pink (like brick dust), and the cap is large, dull brownish gray, streaked, and rimulose. The base of the stalk is set in a large membranous volva. The volva is dull dark brownish gray over the upper part and white below in buttons. There is no annulus. Young gills are whitish.

Edibility Edible and choice; we prefer buttons in which the volva is intact or young caps with whitish gills.

When and where Gregarious on piles of decaying vegetable matter such as sweepings from stables, piles of leaves, rice straw, sugarcane residue, and compost heaps during warm to hot wet weather. It is widely distributed in North America and abundant in suitable habitats—which seem to be few and far between.

181 *Volvariella volvacea*   About one-half natural size

Microscopic features   Spores 7.5–9 × 5–6 μm. Pleurocystidia 52–105 × 7.5–15 (25) μm. Cheilocystidia 30–60 (107) × 8–18 (29) μm.

Observations   This species is cultivated in the Orient and Southeast Asia generally where it is the common mushroom of commerce. Other names for it include Chinese mushroom and paddy straw mushroom. The collection shown here is from Michigan, but the species has been reported from many stations in the South. The specific epithet means having a volva or wrapper.

---

182   *Pluteus cervinus* (Schff. : Fr.) Kumm.
(Deer Mushroom)

Identification marks   The cap is dark gray to grayish brown or occasionally nearly black and darkest over the disc. It is smooth and moist on the surface and has a radishlike taste. The stalk is white or tinged with the color of the cap. When young, the gills are nearly white but they become pink as the spores mature.

Edibility   Edible but soft when cooked; the radishlike taste persists.

When and where   Solitary to clustered on rotting wood or arising from buried wood, usually of broad-leaved trees; spring and summer in the North to early winter in the South; widely distributed in North America.

Microscopic features   Spores 6–7.5 × 4.5–5.5 μm. Pleurocystidia 60–90 × 12–18 μm, walls thickened, hornlike projections present at the apex. Cuticle of cap of radially arranged appressed hyphae.

Observations   It flourishes on dead wood and sawdust piles and is one of the species that "cleans up" the debris left by lumbering operations. *Cervinus* means pertaining to a deer.

182   *Pluteus cervinus*   Four-fifths natural size

---

183   *Pluteus pellitus* (Pers. : Fr.) Kumm.

Identification marks   The disc of the cap is light grayish brown to brown and rimulose, the remainder is white or nearly so. The caps are soft and thin fleshed. There is no volva at the base of the stalk. Both the odor and taste are slightly radishlike.

Edibility   Edible.

When and where   Late summer into fall, on or arising from dead wood (buried presumably in the case of the specimens illustrated here) of hardwood trees; widely distributed.

Microscopic features   Spores 6–7.5 × 4–5 µm. Pleurocystidia 45–75 × 8–17 µm, thick walled, with hornlike projections at the apex. Cuticle of cap of appressed radiating hyphae.

Observations   This is one of the variants in the *P. cervinus* complex and is sometimes considered a variety of *P. cervinus*. The specific epithet means covered with skins.

183   *Pluteus pellitus*   One-half natural size

Rhodophyllaceae

Two very important features of the members of the family are that the gills are attached to the stalk at least in young specimens and the spores are reddish cinnamon to the color of brick dust in deposits. Most species are terrestrial. The spores are thin walled, lack an apical pore, and are angular or grooved as seen under a microscope. The species are difficult to identify on the basis of field characters for the most part, and enough are poisonous to discourage random sampling of unidentified specimens. We include only two species although *Entoloma* as a genus ranks in size with genera like *Lactarius*. The name *Entoloma* means with a fringe and *Rhodophyllus* means with pink gills.

Key to Species

1.   Cap gray, smooth; gills decurrent; whitish amorphous masses of tissue often present near "normal" fruiting bodies . . . .
. . . . 184. *Entoloma abortivum*
1.   Cap brownish violaceous to bluish purple, velvety to fibrillose-scaly; gills not decurrent; not regularly associated with "aborted" fruiting bodies . . . . 185. *E. violaceum*

184　*Entoloma abortivum* (Berk. & Curt.) Donk

Identification marks　Some fruiting bodies look like typical mushrooms with gray dry caps and stalks, grayish gills that are decurrent and become pinkish by maturity, and no veils. Others are whitish to dull pinkish tan fleshy masses of tissue called carpophoroids. They are formed from young "normal" fruiting bodies that were infected by the mycelium of the honey mushroom.

Edibility　Both the normal fruiting bodies and the carpophoroids are edible. Use only those that are firm and show no signs of decay.

When and where　August to December under hardwoods or in mixed woods, often on or near dead wood; widely distributed in eastern North America and to be expected throughout the South.

Microscopic features　Spores 7.5–10 × 5–6.5 µm, angular. Hymenial cystidia absent.

Observations　This species is sporadic in its appearance in the South but is periodically abundant. When we first visited the collection illustrated, only carpophoroids were present, five days later both normal fruiting bodies and carpophoroids were present. Because many species in this genus are difficult to identify in the field, be sure carpophoroids are associated with any "normal" specimens gathered for eating. The specific epithet means abortive.

184　*Entoloma abortivum*　　　　　　　　　　　One-third natural size

185　*Entoloma violaceum* Murr.

Identification marks　The velvety to fibrillose-squamulose cap is bluish purple to brownish violaceous and lighter shades of the same colors are repeated on the stalk. The gills are attached and tinged with dull pink at maturity. The spores are orange cinnamon to vinaceous buff in deposits. Both the odor and taste are mild.

Edibility   We have no data on it; not recommended.

When and where   Scattered on the ground and on decaying wood in deciduous and mixed forests; late summer and fall. It was described from New York and occurs in the Appalachian Mountains.

Microscopic features   Spores 8–9.5 × 6–7 µm, 5–6-sided. Cheilocystidia 25–45 (74) × 7.5–13.5 (20) µm.

Observations   *Violaceus* means violet.

185   *Entoloma violaceum*

About natural size

Bolbitiaceae

Rusty brown to earth brown (deep brown) spores in deposits are about the only features of this family that can be checked without a microscope. The other characters—smooth spores with an apical pore and a cellular cuticle—can be studied only with a microscope. In general, species with a cellular cuticle do not "peel" readily from the margin toward the disc.

This is one of the smaller families of gilled mushrooms. Many species fruit in lawns and gardens and these are among the first mushrooms "collected" by small children in the "grazing stage" (when every object they find goes into their mouth). One species of *Conocybe* (not usually found in lawns) is known to contain amanitin and is potentially dangerous. The two common genera are *Conocybe* and *Agrocybe*. The former name means conic head or cap, the latter means field cap, i.e., a mushroom common in cultivated regions.

Key to Species

1. Spores bright rusty brown in deposits; cap conic; stalk fragile, 1–3 mm thick .... 186. *Conocybe lactea*
1. Not as above: spores deep rusty brown to earth brown; stalk seldom fragile 2
    2. Fruiting on wood; cap yellowish gray to grayish brown; stalk annulate .... 187. *Agrocybe aegerita*
    2. Fruiting on the ground, often in lawns; cap white to yellowish; stalk not annulate .... 188. *Agrocybe retigera*

186 *Conocybe lactea* (Lange) Métrod

Identification marks These slender mushrooms are a common sight in lawns in the summer. The broadly conic cap is light ochraceous tan in the center and pale buff to nearly white at the margin or is white overall when faded. No veils are present. Fruiting bodies often appear in late afternoon, expand by the next morning, and collapse by early afternoon.

Edibility Not recommended; however, it has been sampled by enough young children to indicate that it is not violently poisonous in small quantities.

When and where Solitary to gregarious in lawns, golf courses, and other grassy areas; late spring and summer after warm rains. This is one of the common lawn mushrooms in North America.

Microscopic features Spores 13–16 (18) × 7–9 μm, truncate from an apical pore. Cheilocystidia capitate, soon collapsing.

Observations One species of *Conocybe*, *C. filaris*, is known to contain poisons of the amanitin group (see p. 11). Its cap is light brown, an annulus is present on the stalk, and the spores are 7–9 × 4–5.5 μm. *Lacteus* means milky or milk white.

186 *Conocybe lactea*   Two-thirds natural size

---

187 *Agrocybe aegerita* (Brigantini) Sing.

Identification marks This large, handsome mushroom typically fruits in clusters on living or dead trees. The smooth yellowish gray to grayish brown caps may become cracked in age and be as much as 20 cm broad. The membranous veil persists as a skirtlike annulus. Neither the odor nor the taste are strong.

Edibility   Edible and popular in many parts of the world such as southern Europe and South America. It has not attracted much attention in North America.

When and where   On trees, especially box elders, cottonwoods, poplars, and other members of the maple and willow families; fall and early winter. It is widely distributed in warm-temperate and subtropical regions, but its range in North America is not known. The specimens shown here were found in Mississippi.

Microscopic features   Spores 9–11 × 5–6 (7) μm. Pleurocystidia 30–45 × 6–11 μm. Cheilocystidia 12–22 × 7–12 μm.

Observations   This species might be confused with *Pholiota destruens* but the latter has whitish soft scales on the cap, a narrow ragged annulus, and soft scales on the stalk. *A. aegerita* is reported to be relatively easy to cultivate on blocks of wood. The specific epithet refers to the black poplar of Europe, a common substrate. Information on this species can also be found under the name *A. cylindracea* in some books.

187   *Agrocybe aegerita*   One-half natural size

Agrocybe retigera (Spegazzini) Sing.   **188**

Identification marks   The cap is white to ivory or light yellowish tan and its surface is wrinkled or pitted, definitely uneven. A thin veil may be evident in buttons but little trace of it remains on mature specimens. The base of the stalk is often in the form of a small bulb. Both the odor and taste are mealy.

Edibility   Not recommended; there is little about it to attract the mycophagist. We have no data on it.

When and where   Gregarious to scattered in lawns, fields, and open areas during hot wet weather from spring to fall; common; to be expected in Florida and the coastal plain. Spegazzini, one of South America's great mycologists, described it from Paraguay, and it is widely distributed in the warmer parts of the Western Hemisphere.

Microscopic features   Spores 15–17 (18) × 7.5–9 μm, truncate. Pleurocystidia (20) 45–75 × (9) 21–33 μm. Cheilocystidia 30–45 × 9–20 μm.

Observations   Little has been recorded about its distribution in North America but it is known from the Gulf Coast region. The specific epithet means covered with a net, a reference to the pattern of wrinkles on the cap.

188 *Agrocybe retigera*   Three-fourths natural size

Cortinariaceae

The spores of most species are yellow brown, rusty orange, or rusty brown in deposits, lack an apical pore, and are ornamented to some degree in all but one genus. The cap and stalk do not separate cleanly. When a partial veil is present it is usually cobweblike and called a cortina; only in a few species is it membranous.

Although this is one of the largest families of gilled mushrooms, it is of relatively minor importance to mycophagists. There are a few well-known edible species. The majority, however, remain untested or are known to be poisonous. It is not a safe group with which to experiment.

Key to Genera and Species

1. Fruiting bodies routinely associated with wood 2
1. Fruiting bodies routinely terrestrial 3
    2. Cap glabrous, moist, butterscotch brown; stalk with a membranous annulus; spores ochraceous to rusty brown in deposits . . . . 189. *Galerina marginata*
    2. Cap fibrillose or if glabrous, yellow to yellow brown; stalk with only a few fibrils if any as remnants from a thin veil; spores bright rusty orange to rusty brown in deposit . . (p. 208) *Gymnopilus*
3. Membranous, collarlike annulus present on stalk; cap with a silvery sheen when young . . . . 190. *Rozites caperata*
3. Not as above 4
    4. Young gills distinctly colored (red, violet, lavender, orange, dark brown, or yellow) or if tan to whitish then cap thickly slimy; cortina always present on young specimens . . (p. 211) *Cortinarius*
    4. Young gills pale tan, off-white, or pale grayish tan; cap dry, moist, or thinly viscid 5
5. Cap fibrillose to fibrillose-scaly or if smooth merely moist; gill edges not significantly paler than the faces in young specimens . . . .
. . . . (p. 214) *Inocybe*
5. Cap glabrous and viscid to some degree when fresh; gill edges often paler than the faces when young . . . . (p. 217) *Hebeloma*

206

Galerina marginata (Batsch) Kühner 189

Identification marks A smooth, butterscotch brown, moist cap; light brown gills; and a slender brown stalk with, at least in youth, a thin, bandlike annulus, are important attributes. Both the taste and odor vary from mild to farinaceous.

Edibility Dangerously poisonous; it contains amatoxins (see p. 11).

When and where Gregarious to clustered on rotting wood of both conifers and hardwoods; fall, winter (in warm areas), and spring; widely distributed in North America.

Microscopic features Spores 8–9 (10) × 5–6 µm. Pleurocystidia 37–75 × 9–11 µm.

Observations The complex of species including *G. marginata*, *G. autumnalis*, and *G. venenata* present challenges to the scientist and danger to any mycophagist who believes that all mushrooms that fruit on wood are edible. The three species have much the same appearance and cannot be readily separated in the field. Several collectors have mistaken one of these species for the honey mushroom and suffered acutely as a result. A *galerum* is a helmetlike cap. *Marginatus* means with a border.

189 *Galerina marginata*   Three-fourths natural size

Rozites caperata (Pers. : Fr.) Karst. 190

Identification marks A silvery gray, thin overlay of fibrils covers the cap and persists longest over the center. The surface of the cap is often wrinkled along the radii. A distinct membranous annulus is present on the stalk and a tight, whitish inconspicuous volva encompasses the base. The gills are light grayish tan and wrinkled somewhat like crepe paper.

Edibility Edible and good; one of our favorites.

When and where Scattered to gregarious under hardwoods and in mixed woods, especially with white pine and hemlock; late summer and fall; confined to the mountains in the South, but widely distributed further north in eastern and western North America.

Microscopic features Spores 10.5−14 × 7−9 μm, some with an apical "snout," minutely warty to roughened. Cuticle of cap a zone of more or less globose cells with an overlay of narrow hyphae.

Observations This is not an easy species to learn but is easily recognized once one has become familiar with it. Once learned, it is an excellent addition to a mycophagist's menu. *Rozites* is named for E. Roze (1833−1900), a well-known French mycologist. *Caperatus* means wrinkled.

190   *Rozites caperata*   About one-half natural size

Gymnopilus

The species in this genus have bright ochraceous to orange spores in deposits, nearly always occur on wood, and lack any viscid or gelatinous layers in the cap. Species of *Gymnopilus* and *Pholiota* are often similar in habit and habitat but in the latter genus the spores are often grayish in deposits, are smooth, usually have an apical pore, and the caps are viscid to the touch or have a layer of gelatinous tissue in them. *Gymnopilus* means naked or bald cap.

Key to Species

1. Cap fibrillose-scaly, the scales yellowish brown; taste mild . . . .
. . . . 191. *Gymnopilus fulvosquamulosus*
1. Cap glabrous to slightly fibrillose; taste often bitter 2
   2. Cap seldom over 6 cm broad; stalk slender, lacking conspicuous remains of the veil . . . . 192. *G. liquiritiae*
   2. Cap often over 6 cm broad; stalk fleshy and often with a distinct zone of fibrils . . . . 193. *G. spectabilis*

---

191   *Gymnopilus fulvosquamulosus* Hes.

Identification marks The cap is covered with dry appressed to slightly recurved, fibrillose, dull yellowish brown scales on a brighter yellow background. The thin fibrillose partial veil leaves only a thin evanescent ring on the stalk. The taste of the cap is mild.

Edibility   We have no information on it, but its small size makes it of little interest to the mycophagist; not recommended.

When and where   Little is known of its distribution; it was described from Michigan, reported from the Great Smoky Mountains, and is shown here from a collection found in Mississippi. It fruits on rotting hardwood logs and stumps in the fall.

Microscopic features   Spores 8–9.5 × 5–6 µm, dextrinoid. Cheilocystidia (24) 33–37, apex 4.5–8 µm broad. Pleurocystidia 22–27 × 5–7 µm, inconspicuous.

Observations   This distinctive species is probably more common than reports indicate. *Fulvosquamulosus* means having yellowish brown scales.

191   *Gymnopilus fulvosquamulosus*   About natural size

Gymnopilus liquiritiae (Pers.) Karst.   **192**

Identification marks   The rich reddish brown to dull orange caps are glabrous. In age the gills may be spotted with reddish brown. No veils are present even on buttons. The mycelium at the base of the stalk is yellow to tawny rather than white.

Edibility   Not reported and not recommended.

When and where   It may fruit on either wood of conifers or hardwoods, but in our experience it is more common on hardwoods; it fruits in the summer and fall and is widely distributed in North America.

Microscopic features   Spores 7.5–9 × 4.5–5.5 µm. Hymenial cystidia 20–40 × 3–7 µm, inconspicuous.

Observations   *G. flavidellus* is somewhat similar but it is common on pine debris in the fall and winter and has white to whitish basal mycelium. In *G. liquiritiae* sections of the cap mounted in KOH and examined with a microscope release copious oily material that obscures the anatomical details. The specific epithet means pertaining to licorice, a reference to its flavor.

192 *Gymnopilus liquiritiae*   Two-thirds natural size

193 *Gymnopilus spectabilis*   About one-half natural size

---

193   *Gymnopilus spectabilis* (Weinmann : Fr.) Sing.

Identification marks   These large caps can be 10–20 cm broad at maturity. They are dry and pale yellowish buff to dull brownish orange or brownish yellow. The surface is silky to fibrillose and often breaks up into small appressed fibrillose scales in dry weather. The taste is bitter, the odor mild to fragrant. A yellowish fibrillose veil is present on buttons and persists as a fibrillose, sometimes evanescent, zone on the stalk.

Edibility   Poisonous; it contains psilocybin and psilocin that may cause hallucinations when ingested (see p. 11).

When and where   Gregarious to cespitose on rotting logs and stumps of hardwoods and conifers; late summer until the arrival of cold weather; widely distributed in North America and sporadically abundant.

Microscopic features   Spores 8–10 × 5–6 µm, dextrinoid. Pleurocystidia and cheilocystidia similar, 22–30 (45) × 3–8 µm.

Observations   In Japan this species is called the big laughing mushroom because of its hallucinogenic properties. The specific epithet means notable or remarkable. *G. validipes*, a closely related species, is also hallucinogenic but *G. ventricosus*, which occurs in the West, is not.

Cortinarius

Rusty brown spores in deposit, habit of fruiting on the ground, and cortinate (cobweblike) partial veil are important characters. In many species the gills are brightly colored when young, in fact almost every color can be found on the gills somewhere in this genus. Depending on the species, the cap may be dry, moist, or slimy. *Cortinarius* is the largest genus of gilled mushrooms in the Northern Hemisphere. In this poorly documented genus there are estimated to be at least 800 species, only a few of which are included here. Within the genus, only a few species are of proven edibility, several are poisonous to varying degrees, and most are untested. Some of the toxins isolated from species of *Cortinarius* are slow acting and there are no antidotes for them. Because of these problems, mycophagists are advised to avoid all members of the genus. A cortina, for which the genus is named, is a thin cobwebby veil.

Key to Species

1. Cap dry to moist when young 3
1. Cap slimy when young 2
   2. Young gills pale tan to off-white .... 194. *Cortinarius mucosus*
   2. Young gills violet to lavender .... 195. *C. cylindripes*
3. Fruiting body silvery lavender when young; stalk stout ....
   .... 196. *C. argentatus*
3. Fruiting body deep cinnabar red, deep mahogany red, or purplish red; stalk slender .... 197. *C. marylandensis*

Cortinarius mucosus (Bull. : Fr.) Kickx   **194**

Identification marks   This species occurs under various 2- and 3-needle pines. There is a thick slime veil covering both stalk and cap as is evident in the photograph. The gills are whitish at first, the cap is medium to dark red brown, and the stalk is white and lacks a conspicuous basal bulb.

194   *Cortinarius mucosus*   Two-thirds natural size

Edibility   It is considered to be edible by some European authorities; however, we do not recommend any *Cortinarius* since the species are poorly known in North America and some are poisonous.

When and where   Common under pine during late summer, fall, and winter in the South. It is widely distributed but often misidentified.

Microscopic features   Spores 10.5–13 × 6–7 µm. Cheilocystidia undifferentiated. Cuticle of cap a thick ixocutis. Clamp connections present.

Observations   There are a number of species with most of the characters of *C. mucosus* but that differ in having violet tints on the young gills or on the apex of the stalk.

195   *Cortinarius cylindripes*   Three-fourths natural size

---

195   *Cortinarius cylindripes* Kauffman

Identification marks   Both the cap and stalk are slimy and violet to lavender in young specimens. The cap soon becomes brown but the stalk long retains its violet tones as do the gills. The stalk is equal, not enlarged at the base.

Edibility   Not reported and not recommended.

When and where   Scattered to gregarious under hardwoods particularly oak with an understory of *Vaccinium* (blueberry) and related shrubs; late summer and fall; widely distributed in eastern North America. Ithaca, New York, is the type locality.

Microscopic features   Spores 12–14 × 7–8 µm. Cheilocystidia 19–33 × 15–25 µm. Clamp connections absent. Cuticle of cap an ixolattice.

Observations   *Cylindripes* means cylindric stalk.

Cortinarius argentatus (Pers. : Fr.) Fr.   **196**

Identification marks The overall lilac blue to lilac gray colors may cause novices to identify it as blewits (*Clitocybe nuda*). The cap is silky and slightly tacky at first but soon dry, the disc soon fades to silvery whitish. The base of the stalk is formed into a round to flattened conspicuous bulb. The cortina is whitish violaceous, thin, and often disappears by maturity—a few fibrils of it can be seen in the left-hand specimen. The gills are narrow and grayish lilac until the spores mature.

Edibility Not recommended; several species have much the same aspect and cannot be reliably identified in the field.

When and where Under oak and white pines or under oak; late summer and fall. Its distribution in North America remains to be determined but it is not uncommon in the southern Appalachians.

Microscopic features Spores 7.5–10 × 5–6 µm. Cheilocystidia 25–45 × 4.5–9 µm. Clamp connections present. Cuticle of cap a thin ixocutis.

Observations At its best this is a truly beautiful mushroom. The specific epithet means silver plated, a very appropriate name. The presence of the cortina and the brown spore deposit serve to distinguish it readily from blewits.

196   *Cortinarius argentatus*   About natural size

Cortinarius marylandensis Ammirati & Sm. (nom. prov.)   **197**

Identification marks The deep purplish red to morocco red of the entire fruiting body and its occurrence under hardwoods (especially oak and beech) are two important field characters. The stalk is slender and dry; the cap is dry and nearly smooth. All traces of the thin cortina soon vanish. A drop of KOH on the cap causes it to change to rose then purple.

Edibility   Not recommended; we have very little data on the edibility of any member of the group of species to which this one belongs.

When and where   Scattered to gregarious under hardwoods east of the Mississippi River in the summer and early fall. It is particularly abundant in the bottomland hardwood forests along the Gulf of Mexico.

Microscopic features   Spores 7.5–9 × 4–5 μm, minutely roughened. Hymenial cystidia not observed. Clamp connections present. Cuticle of cap of radially arranged dry hyphae.

Observations   This species is another collective species that shelters several variants under a single name. As the group becomes better understood, changes in the names will no doubt occur. In the older literature it was included under *C. cinnabarinus*. *Marylandensis* refers to Maryland.

197   *Cortinarius marylandensis*   About natural size

Inocybe

The scientific name of this genus means fiber cap or head, and is appropriate since most species have fibrillose to fibrillose-scaly, dry caps. The hymenial cystidia are typically thick walled, and the spores are either smooth or angular to nodulose. Many species of *Inocybe* are facetiously called "LBMs" (little brown mushrooms) because they are difficult to identify to species. In this large genus, several species are known to contain muscarine in quantities sufficient to cause illness in anyone who eats them.

Key to Species

1. Fruiting bodies slender; cap brown, rimose . . . . 198. *Inocybe fastigiata*
1. Not as above 2
    2. Injured areas staining reddish; cap whitish to pale tan or dingy
        ochraceous . . . . 199. *I. pyriodora*
    2. Injured areas not staining red; cap light yellowish brown to light
        honey brown . . . . 200. *I. olpidiocystis*

Inocybe fastigiata (Schff.) Quél.   **198**

Identification marks   The yellow brown, silky fibrillose cap has a central pointed umbo. The cap becomes radially cracked over the surface in age and splits along the margin. The stalk is slender, light in color, and relatively smooth. In this sense the name covers a number of taxa that differ in such features as spore size, odor and taste of the flesh, and color changes on injured areas.

Edibility   Not recommended—a statement we apply to all species of the genus.

When and where   Scattered to gregarious on soil in woods, under shade trees, along trails, under live oaks, and in similar habitats. It is widely distributed in North America and fruits in late summer, fall, and sometimes winter.

Microscopic features   Spores 9–12 (13) × 5.5–6.5 µm, smooth. Pleurocystidia absent. Cheilocystidia 18–60 × 7.5–16 µm.

Observations   The specific epithet refers to a cluster of branches arranged in a parallel, erect manner; in this case small scales formed of groups of hyphae.

198   *Inocybe fastigiata*   Just under natural size

Inocybe pyriodora (Pers. : Fr.) Kumm.   **199**

Identification marks   The strong fragrant, spicy, or sweet odor is an important characteristic of this species. The fibrils on the cap soon become dull yellowish brown and grouped into patches or scales as shown on the specimen at the right in the photograph. In young specimens the cap margin is whitish. Injured areas stain reddish. Specimens of this species are among the largest in the genus.

Edibility   To be regarded as dangerous! *Do not* experiment with species in this genus.

When and where Scattered to gregarious under brush on hard-packed soil, along roads or trails, and around dried up woodland pools; summer and fall; widely distributed in North America.

Microscopic features Spores 7.5−9 × 4.5−6 µm, smooth. Pleurocystidia 40−60 × 19−26 µm, thick walled; cheilocystidia similar to pleurocystidia or smaller and variable in shape.

Observations The specific epithet refers to the fragrant odor—which does not necessarily resemble that of pears.

199 *Inocybe pyriodora*   Two-thirds natural size

200 *Inocybe olpidiocystis*   About two-thirds natural size

200 *Inocybe olpidiocystis* Atk.

Identification marks At first glance these mushrooms appear to be a species of *Hebeloma*. *Inocybe* is, however, the correct genus partly because most of the cystidia on the gills are thick walled and have crystals adhering to them. The fruiting bodies are stout, the cap tan to light brown and tacky to viscid when young and in wet weather but dry otherwise. The stalk is white and not enlarged at the base. Neither the odor nor color changes on injured areas have been reported for this species.

Edibility Not reported and emphatically not recommended.

When and where   The type locality is a lawn in Ithaca, New York; our collection was found in a churchyard in Mississippi. We have no additional information on its distribution but include it because it seems to occur in towns. It fruits in the fall and early winter.

Microscopic features   Spores 9–12 × 5–6 µm, smooth. Pleurocystidia 48–68 × 19–27 µm, thick walled mostly; cheilocystidia similar.

Observations   *Hebeloma sinapizans* may occur in the same habitat but differs in having pronounced cinnamon tones in the cap, a viscid cap, raphanoid odor, and shaggy stalk. An *olpis* seems to have been a type of leather flask used by the Greeks; *cystis* means cyst or cell.

Hebeloma

Members of this genus have subviscid to viscid, dull brown caps; spores that are dingy ochraceous to dull reddish brown in deposits; and gills whose edges are paler than the faces in young specimens. In many species the fruiting bodies have a distinct radishlike odor and taste. Most species are thought to form mycorrhizae with woody plants. Some are poisonous, and none are popular for table use. The genus name means having a fringe in youth, a reference to the fibrillose veil which soon disappears. The white gill edges are caused by abundant cheilocystidia. In periods of high humidity the gill edges may also be beaded with hyaline droplets.

Key to Species

1. Stalk 2–4 mm thick, appressed silky to glabrous . . . .
. . . . 201. *Hebeloma mesophaeum*
1. Stalk 1.5–3 cm thick; white, scaly to squamulose over most of its length . . . . 202. *H. sinapizans*

Hebeloma mesophaeum (Pers.) Quél.   **201**

Identification marks   The thin pallid veil, reddish brown cap that is slimy in wet weather, radishlike odor and taste, stalk that darkens upward from the base, and occurrence under conifers characterize this species. It is extremely variable.

Edibility   To be regarded as mildly poisonous. It is one of the species that small children find and eat—and then get a trip to the hospital "just in case."

When and where   In the South to be expected under pines in the coastal plains and pine and/or spruce at higher elevations, common under conifers generally throughout North America. It fruits during cool weather around Thanksgiving and in early spring, often before morels fruit.

Microscopic features   Spores 7.5–9 (10) × 5–6 µm, apparently smooth. Cheilocystidia 30–48 × 3–7 µm.

Observations   In dry weather the slime is not as conspicuous as it is on the specimens in the photograph. The specific epithet is derived from words meaning middle and dusky.

201 *Hebeloma mesophaeum*   Two-thirds natural size

202 *Hebeloma sinapizans*   About one-half natural size

202 *Hebeloma sinapizans* (Paulet) Gillet

Identification marks Specimens of this species are among the largest in the genus: the caps may be as much as 15 cm broad. Caps vary from deep reddish brown to pinkish tan or ochraceous brown. Both the odor and taste are radishlike. The stalk is relatively thick and covered with coarse white scales formed by the splitting of the surface; no veil is present.

Edibility Not recommended; generally rated as "suspected" or poisonous.

When and where Late summer and fall to early winter; often in arcs or fairy rings and associated with hardwoods or occasionally conifers; widely distributed and common.

Microscopic features Spores 11–14 × 6.5–7.5 µm, surface wrinkled except for the smooth, somewhat pointed apex. Cheilocystidia (37) 45–54 × 4.5–10 µm.

Observations Drops of liquid are often present on young gills of species in this group. The specific epithet is derived from a word for mustard, no doubt a reference to the odor and taste.

Strophariaceae

Field characters common to members of this family include firm to fibrous fruiting bodies; spores that are either pale tan to rusty brown or purplish brown to chocolate brown in deposits; and attached gills. Furthermore, the spores are smooth and have an apical pore, and there is a special kind of cystidium on the gills of many species. These chrysocystidia have an amorphous yellow mass in them when they are mounted in KOH. The cuticle of the cap is typically composed of radially arranged hyphae.

Various ways of defining genera in this family have been proposed. We recognize four genera here in order to use familiar names. *Pholiota* is the easiest to define. Most of its species fruit on wood and have tan to brown spores in deposit (in contrast to *Gymnopilus* in the Cortinariaceae, which has bright rust-colored spores). *Psilocybe*, *Stropharia*, and *Naematoloma* intergrade to the point that they are sometimes placed in a single genus. The traditional distinction is that a membranous annulus is present on the stalk only in *Stropharia* and that chrysocystidia are present in *Naematoloma* and absent from *Psilocybe*.

Key to Species

1. Fruiting on the ground or on dung of herbivores, typically in pastures, lawns, and open areas 2
1. Fruiting on exposed or buried wood 5
   2. Cap pale to dark yellow brown, appressed fibrillose; fruiting on or near manure of herbivores; bruised areas on stalk staining blue . . . . 203. *Psilocybe cubensis*
   2. Not as above 3
3. Stalk long and narrow, lacking a membranous annulus; cap bell-shaped and deep olive brown when young . . . .
   . . . . 204. *Naematoloma ericaeum*
3. Stalk stocky; annulus present or absent; cap convex, white to tan or light brown when young 4
   4. Cap and stalk white or nearly so when young . . . .
      . . . . 205. *Stropharia melanosperma*
   4. Cap tan to light brown; stalk white or nearly so when young
      . . . . 206. *Stropharia hardii*
5. Cap slimy at first and decorated with fibrillose remnants of the veil
   . . . . 207. *Pholiota polychroa*
5. Cap dry to moist, lacking fibrillose veil remnants 6
   6. Fruiting bodies entirely pale greenish yellow at first; gills smoky gray at maturity . . . . 208. *Naematoloma subviride*
   6. Not with both the above features 7
7. Gills smoky gray by maturity 8
7. Gills brown at maturity . . . . 209. *Pholiota prolixa*
   8. Stalk rooting, base deeply inserted in well-rotted wood of hardwoods . . . . 210. *Naematoloma radicosum*
   8. Stalk directly attached to partly rotted wood of conifers, not inserted into the substrate . . . . 211. *Naematoloma capnoides*

203 *Psilocybe cubensis* (Earle) Sing. (Magic Mushroom)

Identification marks The appressed fibrillose caps are slightly viscid, dark to light yellow brown, and for a time decorated with pieces of the broken veil. The ring on the stalk is white and persistent. Young gills are pale tan to cream color but they darken to chocolate brown as the spores mature. Injured areas of the fruiting body, especially the stalk, stain blue.

Edibility Hallucinogenic (see p. 11); in addition headaches, vomiting, and prostration may accompany its use in adults; children may develop high fevers and convulsions; not recommended.

When and where Gregarious on manure of cattle and horses or manured soil, throughout the growing season; common in the coastal plain and Florida; also reported from the islands in the Caribbean and from Mexico south into Argentina. It was described from Cuba.

Microscopic features Spores (11) 13–17 × 8–10 (12) µm. Pleurocystidia collapsing in age, 18–23 × 10–13 µm; chrysocystidia not observed. Cheilocystidia 19–23 × 6–8 µm.

Observations This species has also been known as *Stropharia cubensis*. It is one of the hallucinogenic mushrooms used for "recreational" purposes in this country; however, it is illegal to possess specimens of it.

203 *Psilocybe cubensis*   Two-thirds natural size

204 *Naematoloma ericaeum* (Pers. : Fr.) Sm.

Identification marks The bell-shaped to convex cap is moist and deep olive brown in youth and fades to ochraceous to rusty ochraceous in age. At first the caps are greasy to the touch; they are dry and glazed in age. Before the spores mature the gills are light gray, at maturity they are mottled dark gray and black. The dry stalk is long and narrow; only a few fibrils persist as evidence of the thin veil.

Edibility Not recommended, but we have no data on it.

When and where Gregarious or scattered in grassy damp areas in thin woods and along roads; summer through fall; common throughout the South in the coastal plain and Florida but reported as far north as Ohio.

Microscopic features Spores 11−13.5 × 6.5−8 μm, apex truncate. Pleurocystidia 25−55 × 8−16 μm, as chrysocystidia. Cheilocystidia as chrysocystidia and leptocystidia 19−35 × 6−9 μm.

Observations Some other species of *Naematoloma* with similar field characters also occur in the South but are more likely to be found on sphagnum moss, in low woods, or dried up swamps. The specific epithet means of the heaths.

204   *Naematoloma ericaeum*   About natural size

Stropharia melanosperma (Bull. : Fr.) Quél.   **205**

Identification marks The entire fruiting body is white or nearly so when young. As the spores mature the gills darken to gray and finally violaceous black. The cap may become dingy yellowish over the disc in age. Remnants of the veil persist both as flaps of tissue on the cap margin and as a thin narrow annulus which is often evanescent. No blue stains develop on bruised areas.

Edibility Edible and good, but as with all white mushrooms be sure to check each specimen to avoid including a poisonous *Amanita*.

When and where Gregarious to scattered on manured ground and in pastures; summer and fall after heavy rains; common and widely distributed but most abundant in the coastal plain and Florida.

Microscopic features Spores 10−12 × 6−7.5 μm, truncate. Pleurocystidia as chrysocystidia, 22−30 × 6−11 (17) μm. Cheilocystidia of two types: chrysocystidia and leptocystidia (24−30 × 6−12 μm).

Observations As applied to this species, the specific epithet means black spored. In the field the attached gills distinguish it from species of *Agaricus* in which the gills typically are free.

205 *Stropharia melanosperma*   One-half natural size

206 *Stropharia hardii*   About one-half natural size

206 *Stropharia hardii* Atk.

Identification marks The veil occasionally forms a persistent annulus, otherwise it leaves flaps of tissue on the edge of the cap or disappears entirely—and then one is not likely to suspect the mushroom of being a *Stropharia*. At first the gills are whitish but they become dark brown to chocolate brown at maturity. The stout stature of the specimens illustrated is typical of the species.

Edibility Little is known about its merits as an esculent. What data we have suggest it may cause gastrointestinal upsets when eaten raw but not when cooked; not recommended.

When and where Under hardwoods and more often in lawns, pastures, and open areas; late summer and fall. M. E. Hard first collected it near Chillicothe, Ohio. It is widely distributed and common in eastern North America but most abundant in the Southeast.

Microscopic features Spores 6–7 (9) × 4.5–5 μm. Pleurocystidia and cheilocystidia present as chrysocystidia, 22–30 × 9–12 μm.

Observations The species is named for Miron Elisha Hard (1849–1914), the author of one of the early American guides to wild mushrooms. It is common in lawns in the deep South where it often fruits with species of *Agaricus* and could be confused with them. The gills, however, in the *Stropharia* should be attached, those of species of *Agaricus* free from the stalk.

222

Pholiota polychroa (Berk.) Sm. & Brodie

Identification marks Variability in the color of the cap is the hallmark of this species. In young caps olive, green, or blue green tones predominate; in old ones yellow and orange colors develop. Caps vary from 2 to 10 cm broad but regularly have fibrillose scales over a layer of slime. Veil remnants persist as scales or patches of fibrils on the edge of the cap and on the stalk below the short-lived annulus.

Edibility Not reported and not recommended.

When and where On dead wood (stumps, logs, and limbs) of hardwoods and occasionally conifers; midsummer to late fall; widely distributed east of the Great Plains and more common in the South than elsewhere. It was described from near Waynesville, Ohio.

Microscopic features Spores 6−7.5 × 3.5−5 µm. Pleurocystidia 40−67 × 9−15 µm, not typical chrysocystidia. Cheilocystidia 30−45 × 6−10 µm.

Observations The specific epithet means many colors, this species is the chameleon of the genus.

207 *Pholiota polychroa*   Natural size

Naematoloma subviride (Berk. & Curt.) Sm.

Identification marks The caps are small, only 1−3 (5) cm broad at maturity and both the young gills and stalk are bright greenish yellow. In age the gills become grayish green. The cap is greenish yellow at least on the margin and shades to ochraceous tan in the center. The taste is bitter and unpleasant.

Edibility Not recommended; *N. fasciculare*, a closely related species, has caused gastrointestinal upsets.

When and where In clusters on logs and stumps of conifers, especially pines; fall and winter when the weather is mild. It was described from Cuba and is to be expected in Florida and the coastal plain, as well as at low elevations in the mountains.

Microscopic features Spores (6) 7−8 × 3.5−5.5 µm. Pleurocystidia as chrysocystidia 22−35 (40) × 7−11 µm. Cheilocystidia as leptocystidia 15−52 × 5−8 µm and chrysocystidia.

Observations For anyone familiar with *N. fasciculare* in the North, *N. subviride* looks like stunted specimens of that species. Detailed studies of this group are needed to show if, in fact, a single variable species or two distinct ones are involved. In *N. fasciculare* the caps are 3–8 cm broad and the stalks 5–12 cm long instead of 3–4 cm long as in *N. subviride*. *Subviridis* means somewhat green.

208 *Naematoloma subviride*   Slightly larger than natural size

209 *Pholiota prolixa*   About natural size

209 *Pholiota prolixa* Sm. & Hes.

Identification marks Large clusters of slender fruiting bodies arising from wood of hardwoods are characteristic of this species. The caps are convex to plane and smooth except for a few thin veil remnants near the margin. The gills are crowded. No annulus is formed by the thin veil.

Edibility Not reported.

When and where Gregarious on or arising from dead wood in areas that have been flooded and around woodland pools as they dry up; late summer into winter; known from the Mississippi Valley and the Great Lakes region.

Microscopic features Spores 6–7.5 × 3.5–5.5 µm. Pleurocystidia 33–45 × 8–11 µm, mostly as chrysocystidia.

Observations We suspect this species is more common than reports indicate. Massive fruitings of it occur some years inside the levees along the Mississippi River in Mississippi in late winter and early spring. The specific epithet means stretched or elongated, a reference to the slender, fairly long stalk.

Naematoloma radicosum (Lange) Konrad & Maublanc   **210**

Identification marks The rooting base (a pseudorhiza technically) and the bitter taste together with the honey-colored cap are distinctive. The pseudorhiza is commonly missed by careless collectors.

Edibility We have no data on it but the bitter taste will discourage most collectors from trying it.

When and where Solitary or in clusters on wood of both hardwoods and conifers; summer and fall or early winter. It occurs from the southern Appalachians northward but is seldom abundant.

Microscopic features Spores 5.5–6.5 × 3.5–4.5 µm. Pleurocystidia 26–35 (48) × 10–14 µm, as chrysocystidia. Cheilocystidia 15–34 × 6–9 µm, as leptocystidia.

Observations Specimens of this species are probably mistaken for old specimens of *N. capnoides* by most collectors. That the rooting base is nearly always missed is evident in our photograph. *Radicosus* means with a large root. In a pseudorhiza, however, the structure grows upward from where the mycelium is living and the mushroom forms at its apex; a root, as everyone knows, grows down into the soil.

210 *Naematoloma radicosum*   One-half natural size

Identification marks   Young gills are whitish to grayish, never yellow to greenish, at maturity they are grayish purple brown. The caps vary from light orange brown to dull cinnamon. Both the odor and taste are mild. A thin, fibrillose, buff veil is present on buttons. A few matted fibrils on the stalk are all that remain of it at maturity.

Edibility   Edible; it is a common species late in the season.

When and where   In clusters on decaying wood of conifers, occasionally on hardwoods; fall and early winter, occasionally early spring; widely distributed in North America and particularly abundant where pines are plentiful.

Microscopic features   Spores 6–7.5 × 3.5–4.5 μm. Pleurocystidia 21–30 × (7) 9–14 μm, as chrysocystidia. Cheilocystidia as leptocystidia 16–22 × 6–10 μm and scattered chrysocystidia.

Observations   The dull yellowish brown to cinnamon brown cap and association with conifers separate it from the brick cap (*N. sublateritium*), but in mixed woods both species and some apparent intermediates may be encountered. Occasional bitter specimens occur and should be discarded. The specific epithet means smoky (the color of the mature gills).

211   *Naematoloma capnoides*   About one-half natural size

Coprinaceae

Important characters of this family include the gills which are black, dark gray, or brownish gray at maturity, and the fragile fruiting bodies. The spores have a germ pore and are black, purple brown, fuscous, gray brown, or rarely dull red in deposits. The cuticle of the cap is cellular in many species—one way to ascertain this character without a microscope is to try to peel the cuticle from the margin toward the disc of the cap; if it is of radially arranged hyphae then it often comes off in strips, if cellular it will not.

Of the four genera in this family, *Coprinus* is the most important to mycophagists. In *Panaeolus* and *Anellaria* the mottling of the gills is due to the fact that the spores mature and are discharged in patches rather than in waves or simultaneously. *Anellaria* is further distinguished by the presence of chrysocystidia on the gills. Most members of this family are saprophytic on plant remains or dung of herbivores, a few are parasitic on other mushrooms. The name *Psathyrella* means a small *Psathyra*, which in turn means fragile; *Panaeolus* means variegated; and *Anellaria* refers to the annulus.

Key to Genera and Species

1. Gills becoming black and inky (undergoing autodigestion) as the spores mature and are discharged .... (p. 229) *Coprinus*
1. Gills not undergoing autodigestion 2
  2. Faces of gills mottled at maturity with light and dark gray or black areas 3
  2. Faces of gills not mottled at maturity...... 212. *Psathyrella incerta*
3. Fruiting on dung of herbivores or on manured ground; cap white when young, convex; stalk mostly over 4 mm thick ....
.... 213. *Anellaria sepulchralis*
3. Fruiting in lawns and grassy areas that are not heavily manured; cap dark chocolate brown fading to light brown, conic; stalk 1–3 mm thick .... 214. *Panaeolus acuminatus*

Psathyrella incerta (Pk.) Sm.   212

Identification marks   The cap of this fragile mushroom is dull ochraceous to buff when young and whitish in age. A veil is present on buttons and persists for a short time as flaps of tissue or fibrils on the cap margin, or some-

212   *Psathyrella incerta*   Just under natural size

times as a distinct annulus. The gills are crowded, white at first and purplish gray at maturity. The odor and taste are mild. The stalk is persistently white overall.

Edibility   Edible, but it has little substance. *P. candolleana* is also edible and the two species are not readily distinguished in the field.

When and where   In clusters or scattered around trees and stumps of hardwoods or other woody debris from late spring into fall. Peck described it from New York and it is widely distributed in the forested areas of North America.

Microscopic features   Spores 6–7.5 × 3.5–4 (5) μm, smooth. Pleurocystidia absent. Cheilocystidia 33–40 × 9–12 μm. Cuticle of cap a zone of rounded cells.

Observations   *Incertus* means uncertain or doubtful. *P. candolleana* has spores 7–9 (10) × 4–5 μm and caps that are honey brown before fading to pale tan or whitish.

213   *Anellaria sepulchralis* (Berk.) Sing.

Identification marks   Among the mushrooms commonly found on manured soil and manure piles, this species can be distinguished by its white to yellowish convex cap, white stalk, and lack of any veils. The upper part of the stalk is often beaded with drops of liquid in young specimens. The surface of the cap may become areolate-rimose in dry weather.

Edibility   Edible and excellent according to some reports.

When and where   Gregarious on manure of herbivores or on manured soil; late spring into late fall during warm wet weather; widely distributed in North America and not uncommon in pastures and grazing lands.

Microscopic features   Spores (14) 16–18 (20) × 9–14 μm, smooth, truncate. Pleurocystidia present as chrysocystidia 30–45 × 15–20 μm. Cheilocystidia as scattered chrysocystidia and leptocystidia 19–35 × 10–15 μm. Cuticle of cap of inflated to isodiametric cells.

Observations   Like all common species this one has had a number of names including *Panaeolus phalaenarum* and *P. solidipes*. *Anellaria* means a little ring or collar; the specific epithet means belonging to a tomb; perhaps it was first collected in a graveyard.

213   *Anellaria sepulchralis*   Three-fourths natural size

Identification marks This slender inhabitant of lawns and grassy areas has a dark chocolate brown cap that fades to light brown and becomes wrinkled. The caps are 1–2 cm broad, broadly conic to conic-umbonate, and lack bands of brown and gray characteristic of the haymaker's mushroom (*Psathyrella foenisecii*). The gills are dull gray at first but become mottled dark gray and black as the spores mature. There are no veils.

Edibility Not recommended, little brown mushrooms (LBMs) can seldom be accurately identified solely on field characters.

When and where Gregarious to scattered in open grassy areas in spring and summer; common in the Pacific Northwest and the Southeast, but widely distributed in North America.

Microscopic features Spores 12–15 × 8–11 μm, smooth. Pleurocystidia not observed. Cheilocystidia 25–35 × 7–9 μm. Cuticle of cap of inflated cells.

Observations This species and the haymaker's mushroom have much the same aspect, coloration, and habitat. The haymaker's mushroom, however, usually has a more rounded, zoned cap, and its spores are distinctly ornamented. *Acuminatus* means tapered to a narrow point.

214 *Panaeolus acuminatus*   About natural size

Coprinus

Members of this genus are often called inky caps because of the black inky mess that results from the autodigestion of the caps. The spores are black and give color to the "ink." Some species are edible, but the fruiting bodies continue to mature after they are picked and may become an inky mess in a few hours or overnight. If fresh young specimens, ones in which the gills have not started to turn gray or black, are immersed in water and refrigerated they will keep a few days rather than a few hours. The name of the genus is derived from a term that refers to dung, a common substrate for members of this genus.

Key to Species

1.   Fruiting on dung or heavily manured soil; cap white and powdery when young . . . . 215. *Coprinus semilanatus*
1.   Fruiting on or arising from wood (either buried or exposed) or terrestrial; cap gray, brown, or tan when young 2

2. Young caps pale tan to crust brown . . . . 216. *C. micaceus*
2. Young caps silvery to gray 3
3. Young caps bearing patches of whitish veil remnants as more or less flat scales . . . . 217. *C. americanus*
3. Young caps lacking scaly veil remnants, at most bearing a few silvery, appressed fibrils . . . . 218. *C. atramentarius*

---

215 *Coprinus semilanatus* Pk.

Identification marks   The habitat on old piles of cow dung, the whitish to very pale gray cap with its powdery to somewhat fibrillose veil, and the lack of an annulus are the features of note.

Edibility   Not likely to be collected in sufficient quantity for table use; we have no data on it.

When and where   It is impossible to give meaningful data on its distribution since it is to be expected wherever cattle are pastured. It fruits during warm to hot weather.

Microscopic features   Spores 12–13.5 × 9–11 × 7–8 μm.

Observations   *C. niveus* is a closely related species that also occurs on cow manure but has larger spores (15–19 × 8.5–10 × 11–13 μm), and the cap is typically snow white before the gills have begun to darken. *C. semilanatus* has slightly duller caps, but the difference is slight. The specific epithet means somewhat woolly—referring to the veil which is intermediate between powdery and woolly

215   *Coprinus semilanatus*   Natural size

Coprinus micaceus (Bull. : Fr.) Fr.   **216**
(Mica Cap)

Identification marks   The veil, which appears to be composed of tiny particles of mica (and for which the species is named), is seldom obvious on caps large enough to harvest. The ochraceous tan color of the cap and the habit of fruiting in large clusters around old stumps of hardwood trees are more useful characters.

Edibility   Edible but it has little substance, only specimens in which the gills have not darkened should be used.

When and where   Common and widely distributed; it is one of our common "urban mushrooms," and is often eaten by children in the "grazing stage." It can fruit any time there is appropriate moisture and temperature.

Microscopic features   Spores 7–8 × 4.5–6 µm.

Observations   This, like all common species of mushroom, is variable in some of its "essential" characters. *Apparently* all members of this group are edible, but we have little accurate data for such a statement. Young specimens of the mica cap and those of *Psathyrella incerta* are superficially similar; however, in the latter the gills do not turn black in age and the cap is whitish at maturity.

216   *Coprinus micaceus*   About one-half natural size

Coprinus americanus Patrick (nom. prov.)   **217**

Identification marks   These handsome mushrooms have patches of white to whitish veil remnants scattered over the silvery gray caps when young and fresh. Ochraceous to light brown tones are absent from the caps. In age the gills undergo autodigestion as one would expect of a *Coprinus*.

Edibility Not reported and not recommended because some closely related species are bitter.

When and where In dense clusters to scattered, arising from or attached to dead wood; summer and early fall; widely distributed in eastern North America. In the South, especially along the Gulf coast and in Florida, this species seems to occupy the niche that *C. variegatus* does further north; however, there is considerable overlap in their distributions.

Microscopic features Spores 7.5–9 × 5–6 µm.

Observations *C. variegatus* differs in having narrower spores (7–9 × 4.5–5 µm), more color in the veil remnants, and ochraceous to light brown caps. *Americanus* means, as expected, American.

217 *Coprinus americanus*   About one-quarter natural size

218 *Coprinus atramentarius* (Bull. : Fr.) Fr. (Inky Cap)

Identification marks Typically it fruits in dense clusters arising from buried wood of hardwoods. The caps are gray to nearly black in age, but often have a silvery sheen before autodigestion sets in. The white stalks feature a zone of fibrils near the base which varies from conspicuous to inconspicuous.

Edibility Generally rated edible but it contains coprine (see p. 10). Those who consume alcoholic beverages some hours to a day or more after eating the fungus may experience difficulty.

When and where Worldwide in distribution and common as an "urban" mushroom fruiting in the spring and fall in the Northern Hemisphere but also during the winter where the weather is mild.

Microscopic features Spores 8–11 × 5–6 µm.

Observations As with nearly all such common mushrooms, variation in many characters is to be noted, especially spore size and presence or absence of clamp connections. Clusters of fruiting bodies have been known to break up paved roads, tennis courts, and packed soil in school yards. *Atramentarius* means inky.

218 *Coprinus atramentarius*   About two-thirds natural size

Agaricaceae

A veil is present on young specimens, and when it breaks it typically leaves an annulus on the stalk. At the time the veil breaks, the gills vary from pink or rose to white or nearly so (if the latter they may become pink as they mature). The gills are free of the stalk, and the cap and stalk separate cleanly. By the time the spores are mature, the gills are chocolate brown to blackish brown. The spores are dark chocolate brown to blackish brown in deposits, smooth, and may or may not have a minute apical pore. Only one genus, *Agaricus*, is common in North America and is treated here.

Agaricus is one of the few genera in which good edible species outnumber those that are undesirable for one reason or another. A few are poisonous, but to our knowledge there are no "deadly" poisonous species in North America. Some, however, are better eating than others. Dan Guravich for years has eaten specimens of most species of *Agaricus* with white fruiting bodies that he finds in and near Greenville, Mississippi, and he has reported that very few produced even mild gastrointestinal upsets. The genus reaches its greatest diversity in the United States in the South. The name *Agaricus* has its origin in the word *agaricum*, an ancient name for a polypore that grew on trees. Linnaeus applied the name *Agaricus* to those fungi with gills and the name has been used for a group of such fungi ever since.

Key to Species

1. Cap white when young and fresh and usually remaining so 2
1. Cap colored or with fibrils and/or scales that are some shade of
   gray, brown, or deep purple when young 5

219   *Agaricus xanthodermoides* Murr.

Identification marks In young buttons the gills are whitish then pale pink, not rosy; as they mature they become rosy brown and finally dark chocolate brown. The veil is thin and either clings to the margin of the cap or forms a thin annulus. When dried the specimens are dull gold to yellowish. Yellowish stains develop on bruised areas, otherwise the fruiting bodies are white to faintly tinged buff over the disc of the cap.

Edibility We have one report that it is edible but try it in small quantities if experimenting with it. Some species in the yellow-staining group are mildly poisonous.

219 *Agaricus xanthodermoides*   One-half natural size

When and where   In open grassy areas with scattered trees; March into August; to be expected in Florida, the coastal plain, and Mississippi Delta.

Microscopic features   Spores 6–7.5 × 4–6 µm. Cheilocystidja absent to rare and then 21–24 × 11–13 µm.

Observations   The specific epithet means resembling xanthoderma; in this case *A. xanthoderma* of Europe, a species that readily stains yellow.

220   *Agaricus chlamydopus*   One-third natural size

Agaricus chlamydopus Pk.   **220**

Identification marks   The caps of these stout white mushrooms are softly cottony when fresh and chalky-fragile when dried. The veil sheaths the base of the stalk and leaves a simple flaring annulus reminiscent of a lady's stocking in the act of falling down. The fruiting bodies are large and firm; the caps are often over 15 cm broad.

Edibility   There are few reports on the edibility of this species but in our experience it is edible and good.

When and where   Summer and fall in grassy areas such as golf courses and lawns; it often fruits in arcs or fairy rings. The type collection was gathered in Denver, Colorado; we have seen specimens of this species from Texas and Mississippi but otherwise little is known about its distribution.

Microscopic features   Spores 7–8 (9) × 5.5–6.5 (7.5) µm. Cheilocystidia abundant, 23–66 × 4.5–7 µm.

Observations   The specific epithet is derived from words meaning wearing a cloak and foot. The large size, thick context, soft almost cottony cap, and distinctive veil are important characters.

Agaricus solidipes Pk.   **221**

Identification marks   This species is closely related to the following species, so a check of the spore size may be needed to separate them. The cap is white and may become cracked into scales in age or dry weather, the annulus is thin, and the stalk usually tapers toward the base. The stalk is solid rather than hollow at about the time the veil breaks.

Edibility   Edible, comparable to the meadow mushroom.

When and where   Scattered in grassy areas and lawns; late spring through summer. It was described from specimens collected in Colorado, re-

221 *Agaricus solidipes*   About two-thirds natural size

ported from Florida, and we have examined specimens of it from Texas and Mississippi. Little else is on record about its distribution.

Microscopic features   Spores 8–9 (10.5) × 6–7.5 µm, with a minute apical pore. Cheilocystidia not observed.

Observations   Specimens of *A. solidipes* and *A. campestris* are difficult, if not impossible, to separate from each other using field characters. The most reliable difference is the size of the spores, *A. campestris* has narrower, shorter spores. *Solidipes* means solid stalk.

222 *Agaricus campestris* L. : Fr.
(Pink Bottom)

Identification marks   Among the white species of *Agaricus* with medium-sized fruiting bodies this one can be recognized by the combination of bright rosy pink young gills, thin and often ragged annulus, and stalk that often tapers toward the base. The surface of the cap may be thinly coated with silky fibrils. Dried specimens are cream colored to buff.

Edibility   Edible and one of the most popular mushrooms.

When and where   April to October in the South; in meadows, lawns, and pastures; often forming fairy rings or arcs; common and widely distributed in North America.

Microscopic features   Spores 7–9 × 4.5–5.5 µm, with a small apical pore. Cheilocystidia not observed.

222 *Agaricus campestris*   About three-fourths natural size

Observations   This is a common species "everybody knows." As a result, there is little agreement on the details of the species. Most species of *Agaricus* that are similar in color and stature have been identified as *A. campestris* at one time or another. Fortunately, no cases of serious poisoning seem to have been caused by members of this species complex so the mycophagist may proceed (as usual, at his own risk). The specific epithet means pertaining to fields or plains, and another common name for this species is meadow mushroom.

223   *Agaricus porphyrocephalus*   Just under natural size

Agaricus porphyrocephalus Møller   **223**

Identification marks   Soft, appressed, dull reddish brown to purplish brown fibrillose scales cover the cap. The gills are bright pink about the time the veil breaks and become blackish brown in age. The stalk is short, stout, and bears a simple, scanty, annulus. The context is unchanging to faintly pink and the odor and taste are mild.

Edibility   Edible and good; it is firm and tasty when cooked.

When and where   Little is known about its distribution in North America; our collections were made in southern Mississippi in November and December where it was abundant in lawns and mowed areas.

Microscopic features   Spores 5.5–7 × 3–4 µm. Cheilocystidia rare, 22–45 × 9–17 µm, resembling basidia.

Observations   Compared to *A. argenteus* these fruiting bodies are more squat in stature, the annulus is more fragile, and the spores considerably smaller. The specific epithet means purple headed (or purple capped for a mushroom).

Agaricus argenteus Braendle in Pk.   **224**
(Silvery Agaricus)

Identification marks   The stout stature; silky brownish to grayish brown fibrils on the cap; thin, white, single annulus; and dull reddish brown color of the young gills are important. In age the gills become deep chocolate brown. The caps are 2–5 cm broad.

Edibility   We concur with Braendle who reported it to be edible and good.

When and where   Scattered or in arcs and fairy rings in lawns and open grassy areas; May to November. It was described from Washington, D.C., and is widely distributed in the South.

Microscopic features   Spores 8−11 (13) × 6−7 µm. Cheilocystidia not observed.

Observations   *Argenteus* means silvery. The large spores are useful in confirming identifications of this species.

224   *Agaricus argenteus*   Two-thirds natural size

225   *Agaricus rhoadsii*   About one-half natural size

225   *Agaricus rhoadsii* Murr.

Identification marks   Appressed purplish to purplish gray fibrils cover the cap except for the margin, which is white and often fringed with remnants of the veil. Young gills are pinkish, mature ones are brownish black. The veil is white with yellowish brown flecks on the underside, and forms a persistent, ample annulus. The base of the stalk is bulbous.

Edibility   Not reported.

When and where Solitary to gregarious in open woods; July to September. It was described from Gainesville, Florida, and is expected to be widely distributed at low elevations in the South.

Microscopic features Spores 6−7 × 3.5−5.5 µm. Cheilocystidia as chains of barrel-shaped, subglobose, and clavate cells 6−18 × 6−11 µm.

Observations Arthur S. Rhoads, a well-known plant pathologist, and Murrill collected the type collection and Murrill named the species for Rhoads. The chains of cells on the edges of the gills are distinctive. The white cap margin is unusual.

Agaricus pocillator Murr.   226

Identification marks These slender mushrooms have relatively flat caps that are plane except for an obtuse umbo at maturity. Gray to dark brownish gray soft scales are present on the cap, especially on the disc. The smooth stalk has a bulbous base. The intact veil is whitish to tan and lacks dark droplets on its lower surface. The gills change from white through pink before becoming blackish brown.

Edibility Murrill reported it was "excellent for the table" and we have no reason to question his opinion.

When and where Late spring into fall in thin woods, shaded lawns, and similar habitats; often gregarious. It was described from Gainesville, Florida, and is common in the South, especially in the coastal plains.

Microscopic features Spores 5−6 × 3−3.5 µm. Cheilocystidia 18−37 × 4.5−9 µm, soon collapsing.

Observations *A. placomyces*, a northern species, is similar in appearance but differs in having dark droplets on the underside of the veil, more scales on the cap, and spores 5.5−6 × 3.8−4.5 µm. It is not rare in the southern Appalachians, particularly under fir. *Pocillator* means a cup bearer.

226 *Agaricus pocillator*   About one-half natural size

Agaricus silvaticus Schff.   227

Identification marks The cap is flat in the center and bears small pinkish brown to pinkish cinnamon appressed scales. The gills are bright pink when young and chocolate brown in age. The annulus is ample and pendant, and the stalk is white and smooth below the annulus. We have used a collective species concept here for a superficially similar group of variants.

Edibility   Not recommended, several variants in this group, including the one illustrated here, may cause gastrointestinal upsets of varying severity.

When and where   Some variants fruit under conifers, others, such as the one illustrated, are common in lawns and grassy areas shaded by broad-leaved trees, particularly oak. The group is widely distributed in North America.

Microscopic features   Spores 4–6 × 3–4 µm. Cheilocystidia abundant, 22–30 × 12–22 µm.

Observations   No satisfactory treatment of the North American representatives of this group is available, which makes precise identifications of the variants difficult to impossible. Mycophagists are urged to avoid the group in general. The specific epithet means of the woods.

227   *Agaricus silvaticus*   Three-fourths natural size

Gasteromycetes

The uniting feature of these fungi is that the basidiospores are borne within the fruiting body and are not forcibly discharged from the basidia at maturity. The group is defined more on function than form and includes an almost bewildering variety of fungi. The spore-producing region in a Gasteromycete is called the gleba and is enclosed within the fruiting body until the spores mature or even later. It inspired the name for the group, which means stomach fungi. The important point to remember about this group is that the fruiting body has taken over the function of spore dispersal from the basidia. Methods of spore dispersal are numerous. For example, in *Rhizopogon* and other genera with similar fruiting bodies, the fruiting bodies may decay where they developed or, if they produce strong odors at maturity, squirrels and other small rodents may feed on them and deposit the spores elsewhere in their droppings. In the stinkhorns, the spores are held in a slimy material that smells sickeningly sweet to putrid; insects and other small animals that feed on carrion are attracted by the odor. They eat and/or walk on the slimy gleba and carry it, and thus the spores, away both on and in their bodies. Finally, in the puffballs the gleba is very powdery at maturity and the spores are dispersed by the wind after being liberated from the fruiting body.

Key to Taxa

1. Fruiting body with a putrid to unpleasant odor similar to that of rotten meat and volva present at base of fruiting body .. (p. 252) Phallales
1. Not as above 2
   2. Fruiting body resembling a small potato formed at or just under the surface of the ground; surface covered with fine fibrils and rhizomorphs, staining pink to rose when bruised; gleba not becoming powdery at maturity . . . . 228. *Rhizopogon rubescens*
   2. Not as above 3
3. Fruiting body stalked and stalk extending the length of the fruiting body . . . . 233. *Rhopalogaster transversarium*
3. Not as above, if a stalk is present, it is topped by a spore case 4
   4. Fruiting body with a chambered, spongy to gelatinous stalk; spore case opening by a lobed mouth that is often red in young specimens . . . . (p. 242) *Calostoma*
   4. Not as above 5
5. Immature gleba firm and hard, soon deep blackish purple, fuscous, or dark olive brown . . . . (p. 245) Sclerodermatales
5. Immature gleba firm to similar to soft cheese and white, becoming slimy when darkening and finally powdery . . . . . . (p. 248) Lycoperdales

Identification marks The exterior is creamy white then yellow and finally becomes dusky rose when injured and in age. There are at most only a few rhizomorphs on the surface. The interior is dingy white when young but changes to dull olive, grayish olive, and finally olive brown as it matures.

Edibility Eaten by squirrels, mice, and other small rodents. We lack enough data on their edibility for people to do other than not recommend them.

When and where Solitary or more often gregarious in thin soil (and then forming bumps) or partly exposed, usually under 2- or 3-needle pines; mid-summer until late fall or early winter; widely distributed and common in North America.

Microscopic features Spores (8) 9–10.5 × 3–4 µm, hyaline, smooth.

Observations We have found fruiting bodies up to 6 cm broad. This is probably the most common species of the approximately 150 in the genus known from North America. The name *Rhizopogon* is derived from words that mean root and beard; *rubescens* means becoming red. These fungi form mycorrhizae with woody plants. Usually one has to dig the fruiting bodies out of the duff, and it is often a challenge to locate them. They form an important part of the diet of many small animals.

228 *Rhizopogon rubescens*   Two-thirds natural size

Calostoma

These peculiar puffballs each consist of a chambered, tough-gelatinous, spongy stalk topped by a spore case that has a complex set of walls and that opens by a lobed mouth. In the United States this genus is typically southern and eastern in its distribution. Only four species of *Calostoma* have been reported from this region. The relationship of these puffballs with the other puffballs is a matter of debate. *Calostoma* means beautiful mouth, a reference to the bright red teeth (perhaps better called lips) of some species.

Key to Species

Calostoma cinnabarina Desvaux   229

Identification marks   At first the spore case is enclosed in a gelatinous layer whose inner surface is red. This layer falls away in strips and chunks that resemble tomato seeds to expose the mature spore case. It is red to reddish when first exposed and soon fades to orange or yellow except for the persistently scarlet mouth. The stalk is spongy and dull brownish yellow.

Edibility   Of no significance as an esculent.

When and where   Solitary or more often gregarious in woods and along roads on the ground; spring to early winter; widely distributed east of the Great Plains and south of the Great Lakes.

Microscopic features   Spores 14−28 × 6−11 µm, punctate.

Observations   The specific epithet refers to the color of the juice of the dragon's blood tree which is also the color of cinnabar ore. This is the most common species in the genus in North America. The fruiting bodies resemble cherry tomatoes on stalks.

229   *Calostoma cinnabarina*   About natural size

230 *Calostoma microsporum* Atk.

Identification marks The exoperidium flakes off over the mouth to expose the bright orange red teeth at maturity. The lower part of the mature spore case is studied with small, firm, flat warts which are remnants of the exoperidium. The spore case is 10–20 mm wide and the spongy chambered stalk is 3–7 cm long.

Edibility Of no consequence as an esculent.

When and where Cespitose to gregarious on wet mossy banks, in clay soil, and along roads; fruiting throughout the growing season. It was described from Tennessee and is to be expected generally in the southern Appalachians.

Microscopic features Spores (6) 7–11 × (3.5) 4.5–6.0 µm, ornamentation hard to demonstrate with a light microscope.

Observations *C. ravenelii* and this species have often been confused. The former generally has larger spores (12.5–16.5 × 5.5–7.5 µm), and smaller fruiting bodies—the stalk may be only 0.5–3 (6) cm tall and the spore sack 5–10 mm broad. *C. lutescens* also has a yellowish endoperidium but its exoperidium forms a flared collar around the base of the smooth endoperidium. All three species occur in the southern Appalachians. *Microsporum* means small spored.

230 *Calostoma microsporum*   Just under natural size

Sclerodermatales

One way to distinguish the tough- or hard-skinned puffballs from the true puffballs (Lycoperdales) is to remember that the gleba in the Sclerodermatales is very firm to hard when young *and* deeply colored at the same time. The spore case wall (peridium) often breaks up irregularly or into lobes rather than opening by a pore or flaking away as in many of the Lycoperdales. None of the Sclerodermatales, also called earthballs, are recommended as esculents—when eaten in quantity they are likely to cause gastrointestinal upsets—but small amounts are sometimes used as a seasoning. *Scleroderma* means tough or hard skin; *Pisolithus* means pea stone, a reference to the pealike divisions of the gleba.

Key to Species

1. Gleba divided into pealike compartments separated by dark brown to black gelatinous material .... 231. *Pisolithus tinctorius*
1. Gleba essentially homogeneous .... 232. *Scleroderma polyrhizon*

Pisolithus tinctorius (Micheli : Pers.) Cok. & Couch **231**

Identification marks The interior of the fruiting body is divided into pealike compartments (peridioles) that mature, break open, and liberate untold numbers of spores. The peridioles mature downward from the apex toward the base of the fruiting body. They are separated by a blackish gelatinous material that quickly stains all it touches olive yellow. The form of the fruiting bodies varies from top-shaped with a small sterile base to resembling a giant molar with a branched, columnar base.

Edibility Not recommended.

When and where In woods especially near pines, in fields, along roadsides, in lawns—almost any place with sandy to gravely, well-drained soil; summer, fall, and winter; widely distributed in North America and especially common in the South and West.

Microscopic features Spores (7) 9–12 μm in diameter, ornamented with spines 1–1.5 μm long.

Observations This species is becoming important in forestry because of its ability to form mycorrhizal associations with many types of trees and to aid seedlings in becoming established in poor soils. Some authors now use the name *P. arrhizus* for this species. *Tinctorius* means used in dying, *arrhizus* means without a root, and *pisolithus* means pea stone (referring to the peridioles).

231 *Pisolithus tinctorius*   About one-half natural size

232 *Scleroderma polyrhizon* (Gmelin : Pers.) Pers.

Identification marks   These large (7–13 cm broad), thick-walled earth-balls are a common sight. The surface becomes deeply cracked to scaly long before the peridium opens to expose the chocolate brown gleba. Opened specimens, with the peridium broken into irregular rays, are likely to be found late in the fall or winter.

Edibility   Not recommended; at least some species in this genus cause gastrointestinal upsets when eaten in quantity.

When and where   Solitary or gregarious in open areas, along roads, and in thin deciduous woods and along their borders; late summer into winter; widely distributed in North America and particularly abundant in the South.

Microscopic features   Spores 7–11 μm in diameter (including ornamentation); ornamentation forming at most a broken reticulum, composed of irregular warts to 0.5–0.8 μm tall.

Observations   *S. texense* is similar in appearance and has been reported from Pennsylvania southward and westward in the United States and is common in Mexico. It differs in having many thick walled hyphae in the peridium instead of few or none, more prominent scales on the peridium, and less reticulate spores. *S. geaster* is a later name for *S. polyrhizon*. *Polyrhizon* means many roots.

232   *Scleroderma polyrhizon*   About one-half natural size

233 *Rhopalogaster transversarium* (Bosc) Johnston

Identification marks   Although these fruiting bodies may appear to be buttons of some mushroom, they never develop into "normal" mushrooms. Instead the outer layer of the enlarged portion wears away and exposes the gleba. In the collections we have seen, the gleba was dark greenish brown and slimy—somewhat like the gleba of a stinkhorn. Some authors report that the gleba is powdery in age. In a specimen cut in half from top to bottom, you can see that the stalk extends through the gleba to the tip of the fruiting body.

233 *Rhopalogaster transversarium*   About natural size

Edibility   Not reported; there is little to recommend even testing it.

When and where   Look for it in damp pine woods on tree trunks, fallen limbs, and other debris around pines from summer until the onset of cold weather. Bosc described it from specimens he saw in "la basse Caroline" and the species occurs at least from North Carolina to Florida and west into Mississippi.

Microscopic features   Spores 6–7.5 × 3–3.5 (4.3) μm, smooth, pale honey colored in KOH.

Observations   Specimens that have a powdery white or yellow surface have been attacked by a mold. The relationships of this fungus are debatable. The generic name means club stomach; the specific epithet refers to the stalk that extends completely through the gleba.

Lycoperdales

This is the largest order of puffballs; it includes the true puffballs and earthstars. When young, the gleba in these fungi is white and firm but not hard. At this stage they can be eaten when cooked. Always cut open from top to bottom any puffball destined to be eaten and examine the cut surfaces to be sure that the specimen is a puffball (homogeneous interior) not a mushroom button (interior will show signs of a cap, stalk, and gills), or other fungus. In many species once the spores are ripe, the spore case opens by a distinct pore.

Key to Species

1. Fruiting body large, often over 10 cm broad, peridium (wall of spore case) breaking up and falling way as flakes, mature gleba dull grayish lavender to dull purple . . . . 234. *Calvatia cyathiformis*
1. Not as above: spore case opening by a pore or small opening 2
   2. Outer wall of fruiting body splitting into rays and becoming star-like; spore case set in center of the star . . . .
   . . . . 235. *Geastrum fimbriatum*
   2. Not as above 3
3. Young fruiting bodies covered with more or less erect white spines 4−6 mm long that fall away by maturity; spore case shiny and deep purple brown at maturity . . . . 236. *Lycoperdon pulcherrimum*
3. Young fruiting bodies covered with soft, matted-fibrillose spines and scurf that disappear by maturity; spore case shiny and ochraceous to yellow brown at maturity . . . . 237. *Bovistella radicata*

---

234 *Calvatia cyathiformis* (Bosc) Morgan

Identification marks Cut the specimens open from top to bottom to see the chambered sterile base; it is not easily distinguished from the spore-bearing region in intact specimens. At first the exterior is pale brown and embossed with angular patches of feltlike material or cracked into polygons; by the time the spores are mature the spore case breaks up and falls away over the upper part of the fruiting body. The interior is white and homogeneous above the chambered base in youth but some shade of violet to dull purple by maturity.

Edibility Young specimens in which the gleba (spore-bearing region) is still firm and white are edible and popular; ones in which the interior is changing color or colored may cause gastrointestinal upsets.

When and where Fresh specimens appear in summer and fall; empty sterile bases may be found throughout the year. It fruits in meadows, grassy areas, lawns, and open woods and is widely distributed.

Microscopic features Spores (including ornamentation) 5−6.5 μm in diameter, globose; ornamentation less than 0.5 μm tall. Capillitial threads to 5 μm broad, septate, pitted, and branched.

Observations Several species of *Calvatia* occur in the South; as long as the gleba is firm and white throughout they are all presumed to be edible. Those species in which the gleba is violet to purple at maturity form a complex group in need of study; we merely follow established tradition here in our species concept. The name *Calvatia* means bald or becoming bald; *cyathiformis* means cup shaped, a reference to the shape of the empty sterile base.

234   *Calvatia cyathiformis*   One-half natural size

Geastrum fimbriatum Fr.   **235**
(Fringed Earthstar)

Identification marks   In this earthstar, as in all earthstars, the outer portion of the fruiting body splits into rays that bend back to expose the spore case and, in this species, remain expanded once the fruiting body has opened. The spore case is not stalked, and the mouth is somewhat tubular and not strongly delimited from the rest of the spore case. At maturity the spore case is dark brown.

Edibility   Not recommended; mature fruiting bodies are too powdery to be palatable and one seldom finds many unopened ones.

235 ·*Geastrum fimbriatum*   About natural size

When and where Solitary to gregarious on the ground especially under red cedars and in mixed woods; summer and fall; abundant in the South and widely distributed in eastern North America.

Microscopic features Spores 3−3.5 µm in diameter, globose, minutely ornamented. Capillitial threads 3−6 cm broad, aseptate, solid, unbranched.

Observations Earthstar fruiting bodies may persist for several months or even over winter. *Geastrum* as a genus is of little or no economic importance. The name *Geastrum* means earthstar, *fimbriatus* means fringed.

---

236 *Lycoperdon pulcherrimum* Berk. & Curt.

Identification marks The spines on these puffballs appear formidable but are soft in reality. They are 4−6 mm long and white to ivory at all stages of development. After the spines fall away the deep purplish brown, smooth, shiny spore case is visible. Young fruiting bodies are white and firm throughout, by maturity the gleba is powdery and purple brown. There is a small sterile base.

Edibility Edible when firm and white throughout.

When and where Often solitary, occasionally gregarious; on the ground in woods, brushy places, and open areas. It was described from Pennsylvania and is common in the South and southern Great Plains in the summer and fall.

Microscopic features Spores 4−4.5 µm in diameter, globose, ornamented with spines 0.5−1 µm high; pedicels to 20 µm long attached to spores or broken but visible in mounts of the gleba. Capillitial threads branched, tapered, to 6−7 µm broad; walls pitted, brown.

Observations The specific epithet means beautiful. This species intergrades with *L. americanum* in the North. In it the spines are dark at the tip and the spore case is patterned after the spines are gone. Many species of *Lycoperdon* occur in the South, none are known to be poisonous. *Lycoperdon* means the wind of a wolf.

236 *Lycoperdon pulcherrimum*   Two-thirds natural size

237 *Bovistella radicata*   Just under natural size

Bovistella radicata (Durieu & Mont.) Pat.   **237**

Identification marks   Young specimens are white inside and out; by maturity the gleba is brown and powdery and the walls of the chambers in the sterile base are metallic purplish gray. At first soft lax spines are present on the spore case but they fall off; at maturity the spore case is smooth, shiny, and somewhat bronze. An irregular opening develops through which the spores escape. The base of the fruiting body tapers into a distinct pseudorhiza.

Edibility   When young and white throughout it is edible and said to be good; avoid specimens in which the gleba has started to change color.

When and where   It fruits in the summer and fall on thin sandy soil in open places and on cultivated and disturbed ground and is common in the Southeast but widely distributed in North America. It was described from northern Africa.

Microscopic features   Spores 4.5–5 × 3–4 µm; pedicel 5–12 µm long, persistent. Capillitium in the form of discrete branched units with tapered apices.

Observations   Fruiting bodies may be as much as 15 cm broad. The genus name means a small *Bovista*; *radicatus* means rooting.

Phallales

In this order, the gleba is olive brown to blackish brown, slimy, and malodorous at maturity. The smell has variously been described as sickeningly sweet, like that of a dead animal, or of very rotten meat. The gleba is trod upon and eaten by insects and other small invertebrate animals that are attracted by the odor, and the animals distribute the spores. The gleba has often been cleaned off by the time the fruiting body reaches full size. Fruiting bodies of most stinkhorns and their allies develop rapidly once the "egg" is ready to "hatch," reaching full size in only a few hours. The eggs are gelatinous and resemble a balloon filled with gelatin in texture. In the United States, this group reaches its greatest diversity in the South where between one and two dozen species occur. None of the North American species can be recommended for eating.

Key to Species

1. Fruiting body with a distinct stalk 2
1. Fruiting body consisting of several "arms" united at the tip to form a cage; gleba on their inner surfaces . . . . 238. *Linderia columnata*
    2. Head chambered, the chambers separated by ridges of sterile tissue, flattened-globose in shape . . . .
        . . . . 239. *Simblum sphaerocephalum*
    2. Head pitted, lacking sterile ridges, resembling a large thimble in shape . . . . 240. *Phallus hadriani*

---

238 *Linderia columnata* (Bosc) G. H. Cunningham (Columned Stinkhorn)

Identification marks Two to five spongy, orange red to rosy red columns arise from each egg. They are fused at the apex which is often pointed (the broken point is on the ground in the photograph), but there is no basal stalk. The gleba is spread over the inner surfaces of the columns and is soon removed by insects. Fruiting bodies vary from 5 to 15 cm in height. The volva remains white.

Edibility Poisonous? There are reports that livestock have been poisoned by this species.

When and where Solitary to gregarious often on or near rotting wood or woody debris; summer into winter, but it may be found at other times also. It was described from the Carolinas and may be found from New York southward at lower elevations in the Southeast and westward along the Gulf of Mexico; it also occurs in Hawaii.

Microscopic features Spores 3.5–5 × 2–2.5 μm.

Observations D. H. Linder (1899–1946), former Curator of the Farlow Herbarium at Harvard University is commemorated in the genus name; the specific epithet refers to the columns of the fruiting body. For many years this species was called *Clathrus columnatus*.

238 *Linderia columnata*   One-half natural size

Simblum sphaerocephalum Schlechtendal   **239**
(Chambered Stinkhorn)

Identification marks Each fruiting body develops from a white egg that persists as the volva at the base of the hollow, reddish or orange red stalk. The head resembles a flattened globe and has been described as chambered, pitted with shallow depressions, or reticulate—different ways of evaluating the same features. The gleba is located in the depressions.

Edibility Not recommended.

When and where It fruits during warm weather in lawns, pastures, orchards, and open woods; occasionally abundant. In the United States it oc-

239 *Simblum sphaerocephalum*   One-half natural size

curs at least from New York south throughout the Southeast and west into Kansas and New Mexico. It also occurs in South America.

Microscopic features   Spores 3.5–4.5 × 1.5–2.5 µm.

Observations   *S. texense* is similar in form but is yellow and is reported to have larger spores (7 × 3 µm). *Simblum* means a beehive, a reference to the chambered head; the specific epithet means round head.

---

240   *Phallus hadriani* Ventenat : Pers.

Identification marks   The combination of spongy stalk and thimble-shaped, flat-topped head covered with the slimy gleba is common to all species of *Phallus*. *P. hadriani* can be recognized by the combination of pitted-reticulate head and dusky rose color of the egg and stalk.

Edibility   Not recommended although there are reports that sliced, fried eggs of some species of *Phallus* are nonpoisonous.

When and where   In shaded lawns, hedges, and thin woods; most abundant in summer and fall but fruiting at other times as well. It is widely distributed in North America and not infrequent in the South.

Microscopic features   Spores 3–4.5 × 1.5–2 µm.

Observations   *P. impudicus* lacks the purplish rose colors of *P. hadriani* but is similar in other respects. *P. ravenelii* and *P. rubicundus* have smooth to roughened heads; the former has white to pale pinkish lilac fruiting bodies, the latter orange red ones. All four species have been reported from the South. *Phallus* means a rod; the specific epithet refers to Hadrian.

240   *Phallus hadriani*   About one-half natural size

Appendixes

Collections Illustrated

The following list, by species number, is of the collections illustrated; all collections are deposited in the Herbarium of the University of Michigan, and all photographs were taken by the collector unless otherwise noted. The following abbreviations are used: DG for Dan Guravich, principal collector and photographer, AHS for A. H. Smith, NSW for N. S. Weber, and JAW for J. A. Weber.

1. *Peziza ammophila*, DG 893
2. *Sarcoscypha occidentalis*, DG photo, no number
3. *Urnula craterium*, DG 815, 1168
4. *Gyromitra fastigiata*, AHS photo
5. *Gyromitra caroliniana*, DG 814
6. *Verpa conica*, AHS photo
7. *Morchella semilibera*, AHS photo
8. *Morchella elata*, DG 1177
9. *Morchella crassipes*, DG 1169
10. *Morchella deliciosa*, DG 1174
11. *Cordyceps melolonthae*, DG 1056
12. *Hypomyces lactifluorum*, DG 584
13. *Auricularia auricula*, DG 609
14. *Tremella fuciformis*, DG 400
15. *Tremella concrescens*, DG 1553
16. *Sparassis spathulata*, DG 1271
17. *Hydnopolyporus palmatus*, DG 820
18. *Meripilus giganteus*, DG 135
19. *Ganoderma curtisii*, DG 1519
20. *Pycnoporus sanguineus*, DG 1455
21. *Laetiporus sulphureus*, DG 263
22. *Hydnum repandum*, DG 757
23. *Hericium erinaceus*, DG 314
24. *Clavaria zollingeri*, DG 549
25. *Ramariopsis kunzei*, DG 1269
26. *Clavulina cinerea*, DG 982
27. *Ramaria conjunctipes*, DG 1127
28. *Ramaria murrillii*, DG 1276
29. *Ramaria fennica*, DG 1129
30. *Ramaria subbotrytis*, DG 1124
31. *Craterellus odoratus*, DG 1541
32. *Craterellus fallax*, DG 1080
33. *Cantharellus cinnabarinus*, DG 1539
34. *Cantharellus tubaeformis*, DG 1113
35. *Cantharellus confluens*, DG 832
36. *Cantharellus cibarius*, DG 1224
37. *Pulveroboletus ravenelii*, DG 1069
38. *Boletellus ananas*, DG 666
39. *Boletellus russellii*, DG 854
40. *Boletellus chrysenteroides*, DG 1287
41. *Austroboletus gracilis*, DG 1043
42. *Austroboletus betula*, DG 1004
43. *Austroboletus subflavidus*, DG 1350
44. *Strobilomyces confusus*, DG 971
45. *Strobilomyces dryophilus*, DG 1374
46. *Gyroporus subalbellus*, DG 1360
47. *Gyroporus castaneus*, DG 328
48. *Leccinum albellum*, DG 1216
49. *Leccinum snellii*, DG 1051
50. *Leccinum griseum*, DG 649
51. *Leccinum crocipodium*, DG 803
52. *Leccinum rugosiceps*, DG 1274
53. *Tylopilus conicus*, NSW 4573, JAW photo
54. *Tylopilus rhoadsiae*, DG 1388
55. *Tylopilus balloui*, DG 1200
56. *Tylopilus alboater*, DG 1037
57. *Tylopilus fumosipes*, DG 1214
58. *Tylopilus tabacinus*, DG 1351
59. *Tylopilus plumbeoviolaceus*, DG 1249
60. *Tylopilus indecisus*, DG 1230
61. *Suillus pictus*, DG 1095
62. *Suillus decipiens*, DG 1394
63. *Suillus hirtellus*, DG 1267
64. *Suillus salmonicolor*, DG 162
65. *Suillus placidus*, DG 1046
66. *Suillus brevipes*, DG 1146
67. *Boletus curtisii*, DG 835
68. *Boletus parasiticus*, DG 1112
69. *Boletus hemichrysus*, DG 1232
70. *Boletus piedmontensis*, DG 1132
71. *Boletus frostii*, AHS photo
72. *Boletus flammans*, DG 1033
73. *Boletus bicolor* var. *bicolor*, DG 1031
    Boletus bicolor var. *borealis*, DG 1378
74. *Boletus erythropus*, DG 804
    Boletus hypocarycinus, DG 1493
75. *Boletus pseudosulphureus*, DG 1393
76. *Boletus longicurvipes*, DG 1121
77. *Boletus griseus*, DG 1005
78. *Boletus retipes*, DG 1275
79. *Boletus variipes*, DG 790
80. *Boletus pinophilus*, DG 1229
81. *Boletus affinis*, DG 1030
82. *Boletus viridiflavus*, DG 1272
83. *Boletus caespitosus*, DG 1085
84. *Boletus fraternus*, DG 822
85. *Boletus campestris*, DG 327
86. *Phylloporus rhodoxanthus*, DG 792
87. *Lactarius indigo*, DG 686
88. *Lactarius paradoxus*, DG 771

89. *Lactarius salmoneus*, NSW 4668, JAW photo
90. *Lactarius pseudodeliciosus*, DG photo, no number
91. *Lactarius atroviridis*, NSW 4529, JAW photo
92. *Lactarius luteolus*, AHS photo
93. *Lactarius corrugis*, DG 690
94. *Lactarius volemus*, DG photo, no number
95. *Lactarius piperatus*, DG 641
96. *Lactarius tomentoso-marginatus*, DG 1516
97. *Lactarius allardii*, AHS photo
98. *Lactarius croceus*, DG 506
99. *Lactarius chrysorheus*, DG 1101
100. *Lactarius subvernalis*, DG 687
101. *Lactarius subplinthogalus*, DG 774
102. *Lactarius peckii*, DG 689
103. *Lactarius argillaceifolius*, DG 1012
104. *Lactarius agglutinatus*, DG 1019
105. *Lactarius yazooensis*, DG 784
106. *Lactarius hygrophoroides*, DG 1485
107. *Russula aeruginea*, DG 336
108. *Russula romagnesiana*, DG 922
109. *Russula compacta*, DG 662
110. *Russula subnigricans*, DG 852
111. *Russula balloui*, DG 754
112. *Russula subfoetens*, DG 1291
113. *Russula amoenolens*, DG 1186
114. *Russula pectinatoides*, DG 187
115. *Hygrophorus hypothejus*, DG 890
116. *Hygrophorus roseibrunneus*, DG 950
117. *Hygrocybe nitida*, AHS photo
118. *Hygrocybe marginata*, AHS photo
119. *Limacella illinita*, DG 801
120. *Limacella kauffmanii*, DG photo, no number
121. *Amanita farinosa*, DG 1492
122. *Amanita ceciliae*, DG 1183
123. *Amanita spreta*, DG 183
124. *Amanita vaginata*, DG 1234
125. *Amanita umbonata*, DG 180
126. *Amanita gemmata*, DG 1532
127. *Amanita flavorubens*, DG 1188
128. *Amanita rubescens*, DG 1201
129. *Amanita volvata*, DG 1221
130. *Amanita citrina*, DG 1147
131. *Amanita bisporigera*, DG 1189
132. *Amanita mutabilis*, DG 1526
133. *Amanita muscaria* var. *flavivolvata*, DG 1165
134. *Amanita roseitincta*, DG 1293, DG 585
135. *Amanita cokeri*, DG 395
136. *Amanita thiersii*, DG 1568
137. *Amanita abrupta*, DG 1060
138. *Amanita polypyramis*, DG 1140
139. *Amanita praegraveolens*, DG photo, no number
140. *Amanita onusta*, DG 991
141. *Amanita hesleri*, DG 1489

142. *Chlorophyllum molybdites*, DG 1481
143. *Lepiota sanguiflua*, DG 1364
144. *Lepiota besseyi*, DG 802
145. *Lepiota naucinoides*, DG 935
146. *Lepiota procera*, DG 914
147. *Lepiota humei*, DG 969
148. *Leucocoprinus fragilissimus*, DG 1346
149. *Leucocoprinus luteus*, DG 1373
150. *Leucocoprinus breviramus*, DG 867
151. *Leucocoprinus longistriatus*, DG 1349
152. *Leucocoprinus lilacinogranulosus*, DG 1280
153. *Panus rudis*, DG 318
154. *Asterophora parasitica*, DG 1048
155. *Strobilurus conigenoides*, DG 1026
156. *Oudemansiella radicata*, DG 284
157. *Omphalotus illudens*, DG 772
158. *Flammulina velutipes*, DG 300
159. *Mycena epipterygia* var. *caespitosa*, DG 1138
160. *Armillaria caligata*, NSW 4490, JAW photo
161. *Tricholomopsis formosa*, DG 1233
162. *Pleurotus dryinus*, DG 279
163. *Pleurotus ostreatus*, NSW 4601, JAW photo
164. *Laccaria trullisata*, DG 764
165. *Laccaria ochropurpurea*, DG 1106
166. *Armillariella mellea*, DG 881
167. *Armillariella tabescens*, DG 742
168. *Lentinus crinitis*, DG 1372
169. *Lentinus detonsus*, DG 1334
170. *Tricholoma flavovirens*, DG 1163
171. *Tricholoma acre*, DG 753
172. *Clitocybe clavipes*, DG 983
173. *Clitocybe nuda*, DG 941
174. *Clitocybe dealbata*, DG 872
175. *Collybia iocephala*, DG 977
176. *Marasmius fulvoferrugineus*, DG 398
177. *Marasmius oreades*, DG photo, no collection
178. *Collybia dryophila*, DG 951
179. *Marasmius nigrodiscus*, DG 1490
180. *Volvariella bombycina*, DG 1059
181. *Volvariella volvacea*, AHS 91782
182. *Pluteus cervinus*, DG 1566
183. *Pluteus pellitus*, DG 921
184. *Entoloma abortivum*, DG 1117
185. *Entoloma violaceum*, DG 448
186. *Conocybe lactea*, DG 644
187. *Agrocybe aegerita*, DG 740
188. *Agrocybe retigera*, DG 849
189. *Galerina marginata*, DG 274
190. *Rozites caperata*, DG 1038
191. *Gymnopilus fulvosquamulosus*, DG 320
192. *Gymnopilus liquiritiae*, DG 1495
193. *Gymnopilus spectabilis*, DG 261

194. *Cortinarius mucosus*, DG 888
195. *Cortinarius cylindripes*, DG 752
196. *Cortinarius argentatus*, DG 1103
197. *Cortinarius marylandensis*, DG 1295
198. *Inocybe fastigiata*, DG 908
199. *Inocybe pyriodora*, DG 1002
200. *Inocybe olpidiocystis*, DG 602
201. *Hebeloma mesophaeum*, DG 773
202. *Hebeloma sinapizans*, DG 237
203. *Psilocybe cubensis*, DG 1505
204. *Naematoloma ericaeum*, DG 1154
205. *Stropharia melanosperma*, DG 1182
206. *Stropharia hardii*, DG 932
207. *Pholiota polychroa*, DG 186
208. *Naematoloma subviride*, DG 999
209. *Pholiota prolixa*, DG 811
210. *Naematoloma radicosum*, DG 450
211. *Naematoloma capnoides*, DG 738
212. *Psathyrella incerta*, NSW photo, no collection
213. *Anellaria sepulchralis*, DG 901
214. *Panaeolus acuminatus*, DG 635
215. *Coprinus semilanatus*, DG 302
216. *Coprinus micaceus*, DG 295
217. *Coprinus americanus*, DG 911
218. *Coprinus atramentarius*, DG 217
219. *Agaricus xanthodermoides*, DG 899
220. *Agaricus chlamydopus*, DG 410
221. *Agaricus solidipes*, DG 1482
222. *Agaricus campestris*, DG 847
223. *Agaricus porphyrocephalus*, DG 1151
224. *Agaricus argenteus*, DG 934
225. *Agaricus rhoadsii*, DG 1555
226. *Agaricus pocillator*, DG 415
227. *Agaricus silvaticus*, DG 325
228. *Rhizopogon rubescens*, DG 1135
229. *Calostoma cinnabarina*, DG 989
230. *Calostoma microsporum*, DG 439
231. *Pisolithus tinctorius*, DG 904
232. *Scleroderma polyrhizon*, DG 885
233. *Rhopalogaster transversarium*, NSW 4676, JAW photo
234. *Calvatia cyathiformis*, DG 1290
235. *Geastrum fimbriatum*, DG 1554
236. *Lycoperdon pulcherrimum*, DG 1552
237. *Bovistella radicata*, DG 1536
238. *Linderia columnata*, DG 1148
239. *Simblum sphaerocephalum*, DG 165
240. *Phallus hadriani*, DG 1468

Additional Species That Occur in the South

The following is a list of the species illustrated in the *Mushroom Hunter's Field Guide* (by A. H. Smith and Nancy S. Weber) that have been reported from the South and/or collected by us there. Most of these species are more likely to occur in the upper South and the Appalachian Mountains than in the coastal plains. The species number is as used in that book.

1. *Leotia lubrica*
3. *Sarcoscypha coccinea*
5. *Aleuria aurantia*
6. *Peziza domicilina*
10. *Disciotis venosa*
14. *Morchella angusticeps*
15. *Morchella esculenta*
18. *Helvella crispa*
22. *Gyromitra infula*
24. *Gyromitra gigas* (group)
27. *Tremella foliacea*
28. *Tremella reticulata*
29. *Sparassis radicata*
30. *Polyozellus multiplex*
31. *Fistulina hepatica*
32. *Polyporus squamosus*
35. *Grifola frondosa*
36. *Grifola umbellata*
38. *Hericium ramosum*
42. *Clavicorona pyxidata*
44. *Ramaria stricta*
46. *Ramaria formosa*
49. *Ramaria botrytis*
50. *Clavariadelphus truncatus*
51. *Clavariadelphus pistillaris*
54. *Cantharellus floccosus*
59. *Gyroporus cyanescens*
60. *Boletinellus merulioides*
62. *Tylopilus chromapes*
63. *Tylopilus rubrobrunneus*
67. *Suillus luteus*
70. *Suillus americanus*
71. *Suillus granulatus*
77. *Boletus subvelutipes*
79. *Boletus luridus*
83. *Boletus pallidus*
84. *Boletus chrysenteron*
87. *Strobilomyces floccopus*
91. *Hygrocybe acutoconica*
92. *Camarophyllus pratensis*
93. *Hygrophorus kauffmanii*
94. *Hygrophorus russula*
95. *Hygrophorus sordidus*
96. *Lentinus lepideus*
98. *Catathelasma ventricosa*
102. *Cystoderma granosum*
106. *Clitocybe robusta*
108. *Clitocybe subconnexa*
110. *Hygrophoropsis aurantiaca*
112. *Clitocybula familia*

113. *Collybia acervata*
115. *Collybia maculata*
116. *Collybia butyracea*
121. *Laccaria laccata*
123. *Leucopaxillus laterarius*
125. *Mycena viscosa*
126. *Mycena leaiana*
128. *Mycena galericulata*
130. *Tricholoma pessundatum*
131. *Tricholoma sejunctum*
138. *Lentinellus vulpinus*
139. *Phyllotopsis nidulans*
141. *Pleurotus porrigens*
145. *Amanita brunnescens*
147. *Amanita flavoconia*
149. *Amanita pantherina*
150. *Amanita atkinsoniana*
152. *Amanita fulva*
154. *Amanita phalloides*
160. *Lepiota americana*
161. *Lepiota rubrotincta*
162. *Lepiota clypeolaria*
169. *Agaricus bitorquis*
171. *Agaricus subrufescens*
173. *Agaricus placomyces*
174. *Clitopilus prunulus*
176. *Entoloma salmoneum*
176. *Entoloma murraii*
180. *Cortinarius corrugatus*
181. *Cortinarius violaceus*
182. *Cortinarius semisanguineus*
184. *Cortinarius gentilis*
190. *Galerina autumnalis*
192. *Paxillus involutus*
193. *Paxillus atrotomentosus*
205. *Pholiota squarrosoides*
210. *Naematoloma fasciculare*
212. *Naematoloma sublateritium*
215. *Agrocybe amara*
217. *Agrocybe praecox*
219. *Coprinus comatus*
221. *Coprinus quadrifidus*
225. *Psathyrella foenisecii*
227. *Lactarius deceptivus*
228. *Lactarius subvellereus*
230. *Lactarius neuhoffii*
232. *Lactarius subpurpureus*
236. *Lactarius subserifluus*
237. *Lactarius gerardii*
238. *Lactarius fumosus*

241. *Lactarius subpalustris*
251. *Russula virescens*
252. *Russula crustosa*
253. *Russula variata*
256. *Russula emetica*
260. *Russula brevipes*
262. *Russula albonigra*
263. *Gastrocybe lateritia*
268. *Calvatia gigantea*
269. *Calvatia craniformis*
270. *Lycoperdon marginatum*

271. *Lycoperdon perlatum*
272. *Lycoperdon pyriforme*
273. *Mycenastrum corium*
274. *Geastrum coronatum*
275. *Astreus hygrometricus*
277. *Scleroderma citrinum*
278. *Scleroderma flavidum*
279. *Crucibulum laeve*
281. *Mutinus caninus*
282. *Phallus ravenelii*

Selected Names and Abbreviations of Authorities of Fungal Names

Atk.	G. F. Atkinson, 1854–1918
Beards.	H. C. Beardslee, 1865–1948
Berk.	M. J. Berkeley, 1803–89
Bond.	A. S. Bondartseu, 1877–19??
Bull.	P. Bulliard, 1742–93
Burl.	G. S. Burlingham, 1872–1952
Cok.	W. C. Coker, 1872–1953
Cor.	E. J. H. Corner, 1906–
Curt.	M. A. Curtis, 1808–72
Fr.	E. M. Fries, 1794–1878
Gilb.	E. J. Gilbert, 1888–1954
Hes.	L. R. Hesler, 1888–1977
Hook.	W. J. Hooker, 1785–1865
Jenk.	D. T. Jenkins, 1947–
Karst.	P. A. Karsten, 1834–1917
Kumm.	P. Kummer, 1834–1912
L.	C. Linnaeus, 1707–78
Mont.	J. P. F. C. Montagne, 1784–1866
Murr.	W. A. Murrill, 1869–1957
Pat.	N. Patouillard, 1854–1926
Pers.	C. H. Persoon, 1761–1836
Pk.	C. H. Peck, 1833–1917
Quél.	L. Quélet, 1832–99
Sacc.	P. A. Saccardo, 1845–1921
Schff.	J. C. Schaeffer, 1718–90
Schw.	L. D. de Schweinitz, 1780–1834
Sing.	R. Singer, 1906–
Sm.	A. H. Smith, 1904–
Thrs.	H. D. Thiers, 1919–
Tul.	L. R. Tulasne, 1815–85

Edible Species Recommended for Beginners

Morchella crassipes
Morchella elata
Hydnum repandum
Hericium erinaceus
Laetiporus sulphureus
Cantharellus cibarius
Cantharellus cinnabarinus
Craterellus fallax
Boletus griseus
Boletus variipes
Boletus pinophilus
Suillus brevipes
Suillus pictus
Suillus decipiens
Lactarius indigo
Lactarius paradoxus
Lactarius corrugis
Lactarius volemus
Lactarius hygrophoroides
Flammulina velutipes
Pleurotus ostreatus
Volvariella volvacea
Volvariella bombycina
Agaricus solidipes
Agaricus campestris
Agaricus porphyrocephalus
Calvatia cyathiformis
Bovistella radicata

Glossary

ACRID (taste): causing a biting sensation on the tongue, intensely sharp and burning.

ADNATE (gills): bluntly attached to the stalk.

AGARIC: mycological slang for any gilled mushroom.

AMMONIUM HYDROXIDE: concentrated NH_4OH used in some macrochemical reactions; household ammonia may be used.

AMYLOID: becoming blue, violet, or black when treated with a solution of iodine such as Melzer's reagent (see also inamyloid and dextrinoid).

ANNULUS: the ring of tissue left on the stalk from the breaking of a veil, usually the partial veil.

APICAL PORE: a pore or thin spot at the apex of a spore, see germ pore.

APICULUS: the projection on some basidiospores that bears the scar left when the spore was discharged from the basidium.

APPENDICULATE: margin of a cap that has patches of veil tissue or fibrils hanging from it.

AREOLATE: cracked into more or less hexagonal areas (areolae) much like a dried out mud flat.

ASCO-: prefix used to indicate association with an ascus or ascus-bearing fungus, e.g., Ascomycete, ascospore.

ASCUS (pl. asci): a cell *in* which spores are formed following fusion of two nuclei and division of the resulting nucleus.

ATTACHED (gills): gills that touch the stalk at some point in their development and are attached to it; free is the opposite condition.

BASIDIO-: prefix used to indicate association with a basidium or basidium-bearing fungus, e.g., Basidiomycete, basidiospore.

BASIDIUM (pl. basidia): a cell *on* which spores are formed following fusion of two nuclei and division of the resulting nucleus.

BOLETE: mycological slang for any member of the Boletaceae.

BOLETINOID (of tube mouths in a bolete): elongated along the radii of the cap as in *Suillus pictus*.

BROAD-LEAVED (trees): trees with broad leaves in contrast to the conifers many of which have needlelike leaves.

BUFF (color): pale yellow toned with gray, dingy pale yellow.

BULBOUS (of a stalk): having an oval to abrupt enlargement (bulb) at the base.

BUTTON (mushroom): young mushroom with the veil intact and/or the cap not yet expanded.

CAP: the umbrellalike portion of the fruiting body that bears gills, teeth, tubes, or is smooth on the underside; when a stalk is present, the cap is at the apex of the stalk. *Pileus* is the technical term for it.

CAPILLITIUM: the threadlike elements mixed in with the spores in a ripe puffball. Capillitial means referring to the capillitium.

CELL (of fungi): the living protoplasmic units into which hyphae are divided.

CELLULAR: composed of cells, but often used to refer to a tissue in which the hyphal cells are more or less isodiametric rather than elongated.

CESPITOSE: growing in clusters with the bases of the fruiting bodies attached to each other.

CHEILOCYSTIDIUM: a cystidium on the edge of a gill or tube.

CHLAMYDOSPORE: a spore formed within a cell in a hypha, formation of such spores is not preceeded by nuclear fusion.

CHRYSOCYSTIDIUM: a cystidium whose contents are yellow and coagulated as revived in KOH.

CLAMP CONNECTION: a specialized hyphal branch that occurs on the hyphae of some basidiomycetes.

CLAVATE: club shaped, as applied to a stalk it means thickened evenly to the base.

CLOSE (of gills): a relative term used to indicate the spacing of gills—see crowded.

COLLECTIVE SPECIES: a species whose original description is sufficiently general that more than one taxon can be placed in it.

COMMON NAME: a name in the everyday language of the area and that is widely used.

CONIDIOSPORE: a type of spore formed without nuclear fusion and formed externally, often in great numbers, by a fungus.

CONIFER: a tree that bears cones.

CONIFEROUS: bearing cones, used here to indicate a woods in which the trees are predominantly of this type, e.g., pine, fir, and hemlock (but not used here to refer to cypress).

CONTEXT: the flesh of the cap and stalk.

CORTINA: a cobwebby veil composed of loosely arranged silky fibrils.

CRISTATE: crested, like a cock's comb.

CROWDED (of gills): very close together. Crowded, close, subdistant, and distant are the terms used to describe gill spacing.

CRUSTOSE: crustlike, resembling small flat scabs.

CUP FUNGI: a member of the class Discomycetes of the subdivision Ascomycotina.

CUTICLE: the differentiated surface zone of the cap or stalk. Pileipellis is another term used for it.

CUTIS: a type of cuticle composed of dry interwoven hyphae.

CYSTIDIUM (pl. cystidia): a sterile cell with some feature such as large size, thickened wall, unusual contents, or striking shape. They may be classified by location (pleuro- and cheilocystidia) as well as by content (chryso- and macrocystidia).

DECIDUOUS (of trees): trees that lose their leaves during part of the year or a forest composed of such trees, as used here it refers to certain broad-leaved trees, not cypres trees.

DECURRENT (of gills): extending downward on the stalk.

DEXTRINOID: becoming reddish brown in Melzer's reagent or other iodine-containing solution; see amyloid also.

DICHOTOMOUS: dividing into two.

DISC (of a cap): the central part of the surface extending roughly halfway to the margin.

DISCOMYCETE: a member of the class Discomycetes, an ascomycete with a fleshy fruiting body and an exposed hymenium.

DISTANT (of gills): widely spaced—see also Crowded.

DUFF: the layer of partially decayed plant material on the forest floor.

ECHINATE: with small pointed spines.

ECHINULATE: with very small pointed spines.

EGG: the somewhat egg-shaped stage of a stinkhorn or *Amanita* when the universal veil is still intact.

ENDOPERIDIUM: the inner layer (if two or more are present) in the wall of the Gasteromycete fruiting body.

ESCULENT: as used here, an edible mushroom.

EVANESCENT: vanishing.

EXOPERIDIUM: the outer layer (if two or more are present) in the wall of the Gasteromycete fruiting.body.

FARINACEOUS: having the odor and/or taste of freshly ground wheat, mealy.

FASCICLE: a little bundle.

FERTILE: as used here, capable of forming spores.

FETID: ill smelling, with an odor similar to that of rotting meat.

$FeSO_4$: ferrous sulfate, a solution of 10 percent ferrous sulphate in water used in certain macrochemical tests.

FIBRIL: a thin threadlike strand.

FIBRILLOSE: covered with appressed hairs or threads.

FLOCCOSE: cottony or woolly.

FREE (of gills): not attached to the stalk and not touching it at any time during their development.

FRUITING BODY: the part of the fungous organism that produces and liber-
ates the spores, the mushroom.

FUNGUS: an organism that lacks chlorophyll and reproduces by spores.

FUNGOID: resembling a fungus.

FUNGOUS: adjectival form of fungus.

FURFURACEOUS: roughened with branlike particles.

FUSCOUS: dusky brown.

FUSOID: tapered at both ends, spindle shaped.

GENUS (pl. genera): the first major grouping above the rank of species;
genera are composed of species having certain characteristics in
common.

GERM PORE: a thin spot in the wall of a spore through which a hypha extends when the spore germinates; it is often at the apex of the spore and
thus is called an apical pore.

GILL: the knife-blade-like, radially arranged plates of tissue on the underside of the cap of certain mushrooms. The technical term for them is *lamellae* (sing. lamella).

GLABROUS: bald, without hair.

GLANDULAR DOTS: slightly sticky spots on the stalks of some boletes.

GLEBA: the mass of spores and capillitium (when present) in the Gasteromycetes that is enclosed during all or part of its development.

GLUTEN: viscous material which makes some fungi slimy.

GREGARIOUS: growing in groups but with the specimens separate at the
base, a condition between scattered and cespitose.

HABIT: manner of growth: solitary, scattered, gregarious, or cespitose.

HABITAT: the type of place in which a fungus naturally grows.

HARDWOOD: used here to refer to those trees (mostly deciduous) whose
wood is hard, or to forests of such trees.

HEAD: the differentiated spore-producing portion of some fungous fruiting
bodies, such as the fertile portion of a morel or stinkhorn, not a true cap
as found in the Aphyllophorales and Agaricales.

HETEROMEROUS: having two or more cell types in a tissue as in the
Russulaceae.

HORNS: projection on cystidia usually at the apex.

HYALINE: transparent or translucent.

HYMENIFORM: with the cells arranged in a palisade, as in a hymenium.

HYMENIUM: the spore-bearing layer of tissue on the surface of gills, teeth,
tubes, etc., the spore-bearing cells arranged in a palisade.

HYMENOPHORE: the part of the fruiting body that bears the hymenium.

HYPHA (pl. hyphae): a thread of the vegetative part of the fungous organism and the fruiting bodies.

INAMYLOID: not turning blue or dark rusty orange to red in solutions of
iodine.

INOPERCULATE: lacking an operculum or lid as the asci of some
Discomycetes.

IXO-: a prefix meaning slimy; as used here a type of cuticle in which the
hyphae are embedded in a matrix of slime or gelatinize and produce
slime, e.g., an ixocutis, ixolattice, or ixotrichodermium.

KOH: potassium hydroxide. We use a 2.5 or 3 percent aqueous solution either to check for color changes on the fruiting body or as a mounting
medium.

LAMELLATE: gill-like; *lamella* is the technical term for gill.

LATEX: the liquid, either colorless or colored, that is released when young,
fresh fruiting bodies of some species are cut.

LATTICE: a type of cuticle in which the hyphae are ascending and
interwoven.

LEPTOCYSTIDIUM: any thin-walled cystidium.

LUBRICOUS: having a buttery feel.

MACROCHEMICAL (reaction): any test where a drop of a chemical is put on the surface or context of a fruiting body to see if a change in color results.

MACROCYSTIDIUM: a cystidium that is usually large and has oily contents, typically found in the Russulaceae.

MARGIN: the outermost part near and including the edge either of the cap or a gill.

MELZER'S REAGENT: a solution used to test for the presence of starch composed of 1.5 gms potassium iodide, 0.5 gms iodine, 22 gms chloral hydrate, and 20 gms water. Substances giving a positive, blue-black color are amyloid, those turning rusty red are dextrinoid, those that do not change color are inamyloid.

MEMBRANOUS: resembling a membrane or thin skin.

MIXED WOODS: ones in which both coniferous and deciduous species of trees are common, e.g., a pine-oak woods.

MYCELIUM (pl. mycelia): the mass of hyphae which constitute the vegetative portion of the fungous organism.

MYCOLOGY: the study of fungi.

MYCOPHAGIST: one who eats fungi.

MYCOPHAGY: the practice of eating fungi.

MYCORRHIZAE: an association of a mycelium and the rootlets of a green plant apparently of advantage to both parties.

OCHRACEOUS: dingy yellow to dull brownish yellow.

OPERCULATE: an ascus which opens by a lid, the operculum, to discharge its spores.

ORDER: the first major grouping above the rank of family, e.g., Agaricales.

ORNAMENTATION: sculpturing on the walls of spores such as dots, lines, ridges, and warts.

PEDICEL: a narrow base or stalk.

PERIDIUM: the wall, often layered, of the spore case of a puffball or other Gasteromycete.

PERITHECIUM (pl. perithecia): the flask-shaped structures in some Ascomycetes in which asci develop. The spores from the asci are ultimately forced through the neck and out through the pore of the perithecium.

PILEUS: the technical term for the cap of a mushroom. Pileo- is the combining form of the word, e.g., a pileocystidium is a cystidium on a cap.

PLEUROCYSTIDIUM: a cystidium on the side of a gill.

PLICATE: pleated.

POLYPORE: mycological slang for a woody fungus in the Aphyllophorales with a poroid hymenophore.

PORE: the opening of a tube, a minute hole in the hymenophore on the underside of a cap.

POROID: having pores.

POTASSIUM HYDROXIDE: see KOH.

PRUINOSE: appearing as if lightly frosted or powdered, caused by light reflected by minute particles.

PSEUDORHIZA: a rootlike process whose base is deep in the ground; it actually grows up to the surface and the fruiting body forms at its top.

RAPHANOID: radishlike (taste and/or odor).

RETICULUM: a pattern of lines or ridges connected so as to form a network; a stalk or spore with such a pattern is said to be reticulate.

RHIZOMORPH: a stringlike or thin ropelike aggregation of hyphae.

RIMOSE: cracked.

RING: another term for a membranous annulus.

RUGOSE: wrinkled.

SAPROPHYTE: an organism that derives its nourishment from dead organic matter.

SCATTERED: fruiting scattered over a relatively wide area.

SCIENTIFIC NAME: the technical name of an organism. It is in Latin or words that have been latinized and consists of the name of the genus (capitalized) and the specific epithet (not capitalized), e.g., *Cantharellus cibarius*.

SCURF: small, often branlike scales.

SEPTUM (pl. septa): a crosswall in a hypha that divides one cell from another.

SEXUAL STAGE: the stage in the life cycle of an organism where genetic material from two organisms fuse then divide. Spores are produced either in asci or on basidia as part of the sexual stage of many fungi.

SPHAEROCYSTS: more or less globose cells in the context of a fruiting body; they are characteristic of the Russulaceae.

SPATHULATE: shaped like a spoon with a narrow base and expanded upper portion.

SPAWN: same as mycelium.

SPECIES: populations of individual organisms representing a single kind and having certain characters in common which distinguish those populations from all others.

SPECIFIC EPITHET: the second part of the scientific name of a species; it is not capitalized.

SPINE: small pointed cone.

SPORE: a reproductive unit produced by fungi and some plants that performs, in many cases, the same functions as seeds but is quite different in structure and mode of formation.

SPORE CASE: the portion of the fruiting body of a puffball in which the spores are formed.

SPORE SACK: see spore case.

SQUAMULE: small scale.

SQUAMULOSE: bearing small scales.

STERILE: not producing spores.

STALK: the stemlike portion of a fruiting body, technically called a stipe.

STRIATE: marked with lines, grooves, or ridges.

SUB-: prefix meaning somewhat, a little, almost, or near.

SULCATE: grooved.

TACKY: sticky to the touch.

TAXON (pl. taxa): a taxonomic group of any rank, e.g., order, family, genus, or species.

TERRESTRIAL: growing on the ground.

TOADSTOOL: a name commonly applied to mushrooms, in this country often used to refer to a poisonous mushroom.

TOMENTUM: a covering of soft hair. Tomentose means covered with soft hairs.

TRICHODERMIUM: a type of cuticle composed of more or less erect hyphae of more than one cell in length.

TRUNCATE: abruptly cut off leaving a flat apex (often caused by an apical pore).

TUBE: the basic unit in the hymenophore of a bolete, each tube hangs parallel to the force of gravity and opens by a pore.

TUBE MOUTH: the tissue around the pore of a tube.

TUBERCULATE: bearing tubercles, i.e., wartlike or knoblike growths.

TURF: a cuticle composed of more or less erect elongated cells.

TYPE: the element (a specimen, collection, or illustration) to which the name of a taxon is permanently attached. It is not necessarily typical or representative of the taxon.

UMBO: a raised conic or convex area in the center of a cap.

VARIANT: an unofficial designation or a collection or set of collections differing slightly from the type but for which the user of the term does not want to use a formal designation.

VARIETY: a formal designation for one subdivision of a species.

VEIL: a layer of tissue. Two types are common in mushrooms: the partial veil that extends from the edge of the cap to the stalk in buttons, and the universal veil that envelops the entire young fruiting body.

VENTRICOSE: enlarged or swollen in the middle, e.g., a ventricose stipe.
VERNACULAR NAME: a name in the language of the region.
VISCID: sticky to slimy to the touch, more slimy than tacky.
VOLVA: the remains of the universal veil left around the base of the stalk after the veil has broken.
VOLVAL: pertaining to the volva.
VOLVATE: possessing a volva.

WART: small squatty or pyramidal chunks of tissue on a cap or similar bumps on a spore.

ZONATE: possessing zones.
ZONE: a concentric band of different appearance from the remainder of the surface of a cap.

Selected Books on Mushrooms

References Cited
ATKINSON, G. F., AND H. SCHRENK (1893)
Some Fungi of Blowing Rock, N.C.
  Journal of the Elisha Mitchell Scientific Society 9:95–108.
LANGLOIS, A. B. (1896)
Two Lectures on Botany Delivered at the Catholic Winter School, March 20 and 21, 1896.
  New Orleans: Phillippe's Printery.
LANGLOIS, A. B. (1900)
The Greater Agarics with Special Reference to Their Poisonous and Non-Poisonous Qualities.
  Proceedings of the Louisiana Society of Naturalists 1897–1899. Reprint.
MURRILL, W. A. (1945)
Autobiography.
  Published by the author, Gainesville, Florida.
STEVENS, N. E. (1932)
The Mycological Work of Henry W. Ravenel.
  Isis 18:133–49.
UNDERWOOD, L. M., AND F. S. EARLE (1897)
A Preliminary List of Alabama Fungi.
  Bulletin no. 80 of the Alabama Agricultural Experiment Station of the Agricultural and Mechanical College, Auburn.

History of Mycology
AINSWORTH, G. C. (1976)
Introduction to the History of Mycology.
  Cambridge: Cambridge University Press.
PETERSEN, R. H. (1980)
"*B. & C.*": *The Mycological Association of M. J. Berkeley and M. A. Curtis.*
  Vaduz, Liechtenstein: J. Cramer. Bibliotheca Mycologica 72.
ROGERS, D. P. (1981)
A Brief History of Mycology in North America (Augmented edition).
  Mycological Society of America. Obtainable from the Farlow Herbarium, Harvard University, Cambridge, Mass.

Mushroom Terminology and Nomenclature
MILLER, O. K., JR., AND D. F. FARR (1975)
An Index of the Common Fungi of North America (*Synonomy and Common Names*).
  Vaduz, Liechtenstein: J. Cramer. Bibliotheca Mycologica 44.
SNELL, W. H., AND E. A. DICK (1957)
A Glossary of Mycology.
  Cambridge, Mass.: Harvard University Press.
STERN, W. T. (1966)
Botanical Latin.
  New York: Hafner Publishing Co.
VOSS, E. G., ET AL. (1983)
International Code of Botanical Nomenclature
  Utrecht: Bohn, Scheltema, & Holkema. Regnum Vegetabile 111.

Mushroom Growing
CHANG, S. T., AND W. A. HAYES (EDITORS) (1978)
The Biology and Cultivation of Edible Mushrooms.
  New York: Academic Press.
HARRIS, B. (1976)
Growing Wild Mushrooms.
  Berkeley, Calif.: Wingbow Press.
SINGER, R. (1961)
Mushrooms and Truffles: Botany, Cultivation, and Utilization.
  New York: Interscience Publishers, Inc.

STAMETS, P. AND J. S. CHILTON (1983)
The Mushroom Cultivator, A Practical Guide to Growing Mushrooms at Home.
  Olympia, Wash.: Agarikon Press.

Semitechnical Literature
SMITH, A. H., H. V. SMITH, AND N. S. WEBER (1979)
How to Know the Gilled Mushrooms.
  Dubuque, Iowa: Wm. C. Brown Co., Publishers.
SMITH, A. H., H. V. SMITH, AND N. S. WEBER (1981)
How to Know the Non-Gilled Mushrooms.
  Dubuque, Iowa: Wm. C. Brown Co., Publishers.

Technical Literature
BARONI, T. J. (1981)
A Revision of the Genus Rhodocybe Maire (Agaricales).
  Vaduz, Liechtenstein: J. Cramer. Beihefte zur Nova Hedwigia 67.
BIGELOW, H. E. (1982)
North American Species of Clitocybe, Pt. 1.
  Vaduz, Liechtenstein: J. Cramer. Beihefte zur Nova Hedwigia 72.
COKER, W. C., AND A. H. BEERS (1974)
The Boleti of North Carolina.
  New York: Dover Publications, Inc. Reprint of 1943 edition.
COKER, W. C., AND A. H. BEERS (1951)
The Stipitate Hydnums of the Eastern United States.
  Chapel Hill, N.C.: The University of North Carolina Press.
COKER, W. C., AND J. N. COUCH (1974)
The Gasteromycetes of the Eastern United States and Canada.
  New York: Dover Publications, Inc. Reprint of the 1928 edition.
CORNER, E. J. H. (1950)
A Monograph of Clavaria and Allied Genera.
  London: Oxford University Press.
CORNER, E. J. H. (1970)
Supplement to "A Monograph of Clavaria and Allied Genera."
  Lehre, Germany: J. Cramer.
GUZMAN, G. 1983.
The Genus Psilocybe.
  Vaduz, Liechtenstein: J. Cramer. Beihefte zur Nova Hedwigia 74.
HALLING, R. E. 1983.
The Genus Collybia (Agaricales) in the Northeastern United States and Adjacent Canada.
  Braunschweig, Germany: J. Cramer. Mycologia Memoir No. 8.
HESLER, L. R. (1967)
Entoloma in Southeastern North America.
  Lehre, Germany: J. Cramer. Beihefte zur Nova Hedwigia 23.
HESLER, L. R., AND A. H. SMITH (1963)
North American Species of Hygrophorus.
  Knoxville, Tenn.: The University of Tennessee Press.
HESLER, L. R., AND A. H. SMITH (1979)
North American Species of Lactarius.
  Ann Arbor, Mich.: The University of Michigan Press.
JENKINS, D. T. (1977)
A Taxonomic and Nomenclatural Study of the Genus Amanita Section Amanita for North America.
  Vaduz, Liechtenstein: J. Cramer. Bibliotheca Mycologica 57.
OVERHOLTS, L. O. (1953)
The Polyporaceae of the United States, Alaska and Canada.
  Ann Arbor, Mich.: The University of Michigan Press.
PETERSEN, R. H. (1975)
Ramaria Subgenus Lentoramaria with Emphasis on North American Taxa.
  Vaduz, Liechtenstein: J. Cramer. Bibliotheca Mycologica 43.
PETERSEN, R. H. (1981)
Ramaria Subgenus Echinoramaria.
  Vaduz, Liechtenstein: J. Cramer. Bibliotheca Mycologica 79.

SINGER, R. (1977)
The Boletineae of Florida.
  Vaduz, Liechtenstein: J. Cramer. Bibliotheca Mycologica 58.
SINGER, R. (1975)
The Agaricales in Modern Taxonomy.
  Vaduz, Liechtenstein: J. Cramer.
SMITH, A. H., AND H. D. THIERS (1971)
The Boletes of Michigan.
  Ann Arbor, Mich.: The University of Michigan Press.
SNELL, W. H., AND E. A. DICK (1970)
The Boleti of Northeastern North America.
  Lehre, Germany: J. Cramer.

Mushroom Cookery
GRIGSON, J. (1975)
The Mushroom Feast.
  New York: Alfred A. Knopf, Inc.
MYCOLOGICAL SOCIETY OF SAN FRANCISCO, INC. (1963)
Kitchen Magic with Mushrooms.
  Berkeley, Calif.: Mycological Society of San Francisco, Inc.
PUGET SOUND MYCOLOGICAL SOCIETY (P. SHIOSAKI, EDITOR) (1969)
Oft Told Mushroom Recipes.
  Seattle, Wash.: Puget Sound Mycological Society. Republished by Pacific
  Search of Seattle, Wash., as *Wild Mushroom Recipes.*

Mushroom Poisoning
LINCOFF, G., AND D. H. MITCHEL, M.D. (1977)
Toxic and Hallucinogenic Mushroom Poisoning.
  New York: Van Nostrand Reinhold Co.
RUMACK, B. H., AND E. SALZMAN (1978)
Mushroom Poisoning: Diagnosis and Treatment.
  West Palm Beach, Fla.: CRC Press Inc.

Index

Numbers in boldface are species numbers; those in italics indicate pages on which descriptions can be found.

Nancy Smith Weber is an adjunct research investigator at the University of Michigan Herbarium.

Alexander H. Smith, her father, is professor emeritus of botany at the University of Michigan. Smith has been investigating and classifying mushroom species since 1929 and is considered the leading expert in mushroom identification.

Dan Guravich, a resident of Mississippi, is a photographer and has published numerous articles and books on nature.

Also available

The Mushroom Hunter's Field Guide
by Alexander H. Smith and Nancy Smith Weber

A Field Guide to Western Mushrooms
by Alexander H. Smith